THE
GENIUS
OF
TREES

THE
GENIUS
OF
TREES

How They
Mastered the Elements
and Shaped the World

HARRIET RIX

CROWN
NEW YORK

CROWN
An imprint of the Crown Publishing Group
A division of Penguin Random House LLC
1745 Broadway
New York, NY 10019
crownpublishing.com
penguinrandomhouse.com

Originally published in the United Kingdom by the Bodley Head, an imprint of Vintage, a division of Penguin Random House UK, London.

From *Revolt Against the Sun: The Selected Poetry of Nazik al-Mala'ika: A Bilingual Reader*, translated by Emily Drumsta, reprinted by kind permission of Westbourne Publishers Limited.

From *Basti* by Intizar Husain, translated by Frances Pritchett, reprinted by kind permission of the *New York Review of Books*.

From *Memoirs* by Pablo Neruda, translated from the Spanish by Hardie St. Martin, reprinted by kind permission of Agencia Balcells.

Every effort has been made to trace and contact copyright holders. The publishers will be pleased to correct any mistakes or omissions in future editions.

Photographs courtesy of the author and Martyn Rix, except for fossilized *Archeopteris* root, courtesy of William Stein.

Library of Congress Cataloging-in-Publication Data is available upon request.

ISBN 978-0-593-73551-0
Ebook ISBN 978-0-593-73552-7

Manufactured in the United States of America

9 8 7 6 5 4 3 2 1

First US Edition

The authorized representative in the EU for product safety and compliance is Penguin Random House Ireland, Morrison Chambers, 32 Nassau Street, Dublin D02 YH68, Ireland, https://eu-contact.penguin.ie.

For my mother and the Goatchers, who grew countless trees

Contents

Introduction

There is a power that has been since all eternity, and that force
and potentiality is "viriditas," the greening.

Hildegard of Bingen

I've never been good at saying no. As a result, I found myself in
Iraq for the first time in June 2014, staying with Assyrian friends
in the town of Amedi, an ancient and magnificent walled town on
a mesa—a flat-topped hill bounded by steep escarpments—just
south of the mountains dividing Iraq and Turkey. It was a feast day
to celebrate Noah landing the ark safely on a mountaintop, so the
whole extended family came together to eat *pacha*, a sort of Assyrian haggis, uncomfortably and very obviously stomach. The talk
was troubled—ISIS was entering Mosul, only 50 miles away—but
just as we started to eat, a breath of jollity arrived: the son of the
house, who had been unsuccessfully hunting wild boar. There are
lots of them in the mountains, he said, they eat the acorns. I was na-
ively amazed—I had never really thought about trees in connection
with Iraq.

That evening we wandered through the oak trees below the
escarpment. There seemed to be many different species and different combinations of leaf shape and bark texture, branch angle, and
acorn size. One big tree had a thick hollow trunk, charred inside
but still living, with long, smooth dark-green leaves and covered in
dark-brown spherical oak galls. These, my host told me, were the
reason Amedi had been rich in the twelfth century: the galls contained chemicals, tannins, that the tree used to defend itself against
insects, but which humans harvested as the key ingredient in early

ink.* Another tree, short and with many stems because it had been coppiced, had small, spiky, silvery-green leaves covered with down. Another, the trunk draped over a large stone as if there was no need for earth, had enormous, sharply serrated leaves, and bunches of Velcro-like hooks hanging down. In the autumn, I was told, they would develop into enormous acorns that humans would eat. All very different from the European oaks I had grown up with in Devon. But they made me feel at home.

When I got back to England I went to stay with my dendrologist grandmother and was put firmly in my place. Not only did she know of people eating acorns in the US and Greece, but she had done it herself in the war and found them not bad, she said, just rather bitter (those tannins again). She was interested to hear about the trees I'd seen. Oak species vary considerably, she explained, and have been growing back into their old haunts since being driven into retreat by the last ice age. Oak trees now stretched almost continuously right around the globe, she went on, in a belt of interrelated species. They dominated forests from North America to the Zagros Mountains in Iraq, and the ones I'd seen were probably trying out the best leaf-shapes for the areas they hadn't moved back to so far. *So far?* She said it as though more than 12,000 years wasn't long at all.

When modern humans evolved about 40,000 years ago, there were an estimated 6 trillion trees on the planet. By the time we appeared on the scene, trees had already altered the planet's air, changed the flow of water, used fire as a tool, and built relationships with the plants and animals around them. For almost 400 million years, trees have been some of the largest organisms on dry land, physically blocking airflows with their branches, channeling waterflows with their roots, and acting as architects for other segments of nature—a mosaic of microhabitats.

---

* Amedi was a place of scholarship and had also had one of the most famous libraries in the region. My host was sparing my feelings—he did not mention that it had been destroyed in wanton violence by British troops in 1919.

We can see this above the ground when we take the time to observe closely: a tree or leaf is a condensation of the place it comes from and bears the marks of its experiences. The leaf of a ginkgo, for example, has veins optimized 385 million years ago, a broad fan-shape that nearly led to its extinction, and the pigments of a changing ozone layer. It is harder to imagine the complexity of a tree below ground, where trees are blind explorers, guided by fungi and bacteria but vulnerable to them, carrying an internal compass of gravity-sensing proteins, their only lodestar the center of the earth.

This book is the story of the agency of trees, a story that—amazingly—no one has told before. How trees communicate and cooperate with one another, what trees do for humans: all this we have been told. But what have trees done simply through the accidents of their evolution, and why? This is the story of how tree-ish trees are, and how, by being tree-ish, they have woven the world into a place of great beauty and extraordinary variety. It is also the story of how we both under- and overestimate trees, and a plea to distinguish rootedness from inactivity and subtlety from simplicity. But while this is a story of science, it's also about taking the science and imagining beyond it, for early in its evolution our very imagination was itself shaped by trees.

It's easy to take our senses and technological genius for granted and forget how ignorant we might be of the very different, embodied genius of these vast organisms. For a start we can try using those senses better than we usually do. Clothing the physical form of a tree in dazzling, virtuosic polyphony is its chemical impact, and here our senses of smell and taste are our way in. Have you ever wondered why bay leaves smell so fragrant and spicy? Why monkey-puzzle bark smells of cucumber? Why a cedarwood coffer smells warm and soft? Scientists can provide names—polyphenols, sesquiterpenes, esters—but they'll be hesitant to give reasons for the chemical behind each scent. To say that a particular scent is to repel herbivores or prevent rot is to provide a tiny sliver of the reason that a tree would create its own chemical cocktails.★

In the same way, a nutritionist can give you percentages and

★ In her book *Good Nature* (Bloomsbury, 2024), Kathy Willis points out that human noses—although not as acute as dogs'—can detect *one trillion* different smells.

chemical formulae for the vitamin C in an orange, or the health benefits of protein in açai, or a picture of improved skin from the fats in an avocado, but the complete story of their construction is that of 20 million years or more of tweaks and biochemical cycles and patience deep within a tree. After all, trees start with the basics: air, water, salts, and sunlight. Even AI is unlikely to find enough information to tell the story fully.* Humans have a tendency to underestimate trees: to assume that if we plant one it will grow, that if we cut one down we can simply plant another. Because trees don't move it is easy to forget that chemically they can run rings round us while we sleep, that they are moving the earth under our feet and shifting the colors of the sky over our head, not just by shading out the sun, but by taking out carbon dioxide and releasing oxygen. Easy to forget, too, that tiny amounts of nitrogen we can't see can unbalance and destroy an oak tree with a trunk thicker than a car that was planted when the Welsh and English were still at war.

We share one world with trees and one need for survival. Over the years of their evolution, trees have become not less but more complicated. When the atmosphere became inhospitable or the earth became too dry, their response was to diversify—to change form to an astounding extent—in order to continue to fit their ecosystem like a glove. The flexibility that we assume in animals, because proteins are inherently more mutable than carbohydrate, is better seen in the microscopic interactions between trees and their environment. Where one tree species burned or starved or was blown away, hundreds more came to fill their place.

What is a tree? And how and why did trees appear on earth? Most of us can look and know: leaf, branch, trunk, tall, rough or smooth bark. The technical definition is a woody plant with one erect perennial stem, its trunk at least 3 inches in diameter at a point 4.5 feet above the ground, a formed crown of foliage, and a mature height of

---

* We have made some progress in understanding the molecules that trees have created: the mass spectrometer can be used to tease out the weight and structure of molecules, x-ray crystallography can reveal complex protein structures, and fluorescent proteins can show some chemical movements happening in real-time.

at least 13 feet, although as with all definitions this is just a starting point.

Bamboo is excluded, no matter how tall, on the grounds of its stem structure—it's a grass. A bonsai is included, however tiny, because it exhibits the chemistry common to all trees, and its miniature leaves and flowers conform to a tree's outward shape.

Why do trees share the attributes they do? Mostly it's a result of the life they lead. Trees are successful because they use sunlight and carbon dioxide to make carbon backbones, in the form of sugar, that provide energy for the rest of life on earth. This is photosynthesis, a process that makes plants both eaten and indispensable. Sunlight and carbon dioxide have both been plentiful for billions of years, and trees inherited the fundamental chemical components to deal with them from their single-celled ancestors. DNA was one essential component. Its stable double helix—the founding structure of life—evolved 4 billion years ago, but its combinations are still typing out the replicating code that allows for life's continuity and development. Chlorophyll evolved in bacteria living in the sea 3.5 billion years ago, enabling sunlight to be captured by trees in the form of chemical electrical energy. Rubisco, the most common enzyme on earth, comprising almost 50 percent of the protein in leaves, evolved 2.4 billion years ago, and enabled photosynthesizing bacteria (called cyanobacteria because they appeared blue) to store the energy they were capturing from the sun by forcing carbon dioxide molecules to form a carbon-carbon bond and enter the stable chain of life as glucose—which can in turn be consumed for energy by other life-forms.*

And about 2 billion years ago, cyanobacteria found a way of spinning many glucose chains together to make cellulose, which allowed them to create a structure and build cell walls. When the first plant cells found themselves beached on land they were burned up over and over again, until finally a free-radical reaction produced a molecule that offered at least some sun-protection and soon became lignin, the woody part of trees.

---

*An enzyme is a biological catalyst: a protein that acts to push two chemicals together in a correct configuration that enables them to react.

So far, so chemical; but as more and more green cells came together, structure became all important. The combination of lignin and cellulose was indispensable in the development of trunks, wicking water upward and providing a casing of Kevlar strength against the decay induced by bacteria and fungi.*

Structure was necessary for efficiency and protection. How this resulted in a type of organism that can grow 380 feet tall, more than twice the height of the Leaning Tower of Pisa, is the truly extraordinary question. To understand how these superstructures formed and diversified we must follow tree evolution down uncertain roads of which only faint traces remain, in fossil sites in French mines and Chinese valleys and German quarries.

Going back so far in time leads us to a strange, apparently accelerated world, in which continents drift around like rubber ducks, bumping into one another. Geological periods are usually defined by a big extinction event in which the world abruptly changes and only a small portion of life on earth carries on. In the Silurian period, for example, 440 million years ago, fungi rather than trees towered over the swamps of the great supercontinent of Gondwana. Draped among them were lichens, strange organisms that kept cooperation between their single cells of algae and fungi loosely structured in order to exploit the violent new environment of the land. But the first land plants with structure and a vascular system were already developing; they grew upward as well as horizontally.

In Lindlar in Germany, fossils indicate plants growing upward in a leafy spike, tufty shrubs about 3 feet long that swirled like millipedes over the forest floor. By the Devonian period about 383 million years ago, trees had developed woody trunks, strengthened by lignin, and cooperated and fitted together as forest much as they do today, while fungi had sunk back into the ground.

There is a recent school of thought that says trees endured and flourished in the Devonian because they had started to shape the abi-

---

*In a book of 80,000 words there is an enormous amount of trapped energy: a library of borrowed tree stability. If a tree is twenty when it is cut and pulped, it would have taken about four months' continuous work for that tree to produce the biomass for one book.

otic environment—water, soil, air—around themselves. Dry land was still a relatively new environment for life and a destructively chaotic one. Typhoons, electrical storms, shifting sediment, bare rock: it was not the gentle, fertile land we know today until trees got to work on it. The best example we have of a fossil forest that can be reconstructed was found at Gilboa in upstate New York in the 1920s. There, rooted fossil tree trunks 3 feet across have been matched up with fossil trunks more than 26 feet tall that end in a fan arrangement of fronds.

The fossil site shows that leaf litter was thick and decayed slowly. The forerunners of centipedes and spiders flourished. Shrubs grew underneath the trees and entwined with their roots. Connections, root grafts, and inosculations—literally "kissings," where trees grow into and merge with one another—were common. The picture of the swampy forest floor we can see from these fossils couldn't be less like a modern plantation, where trees stand upright in rows, but it was productive at a level modern foresters would envy: photosynthesizing fast and locking down lots of carbon dioxide out of the atmosphere, they forced water into clearly defined channels—the beginnings of what we now call rivers.

In 2019, a new discovery added another dimension to our understanding of the impact of trees on the environment 386 million years ago. Excavations in the Cairo quarry near Gilboa in Albany, New York, revealed an enormous 36-foot-wide root system. It belonged to a tree called *Archaeopteris,* which was weathering rocks at a rate no one had expected, in unexpectedly dry conditions. The process was already under way, but trees industrialized it, breaking up rock to make earth and sending sand particles down into the sea, where they reacted with acidic water, locking down dramatic levels of carbon dioxide.

These trees, giant clubmosses and tree ferns, spread widely during the Devonian period and had some of the impact we would lump together today under ecosystem services or natural capital. Flood plains were stabilized by tree-like plants and their roots, and meandering rivers appeared, which were blocked by log jams of clubmoss trunks that looked almost like beaver dams. Trees acted as shelter belts, stabilizing air movements as winds roaring in from the ocean met with resistance in a form that absorbed and diffused the energy. Rain that fell on these coastal forests reevaporated and fell again as rain further inland, reducing the dry desert area of the vast continents and allowing trees to penetrate further into the interior.

While we have only a few rare complete fossils of the trees themselves, the distinctive spores of each species, produced in their millions, are possible to follow across the continents in layers of rock. The spore record reveals huge changes in the species distribution and spread of trees across the world in the first 50 million years of their development. By the Carboniferous period 300 million years ago, trees had spread over so much of the globe and locked down so much carbon that they altered the atmosphere completely. Levels of oxygen went rocketing up from 15 percent to 35 percent, with a correspondingly massive impact on life on earth. Animals became gigantic, plants struggled to adapt, and oxygen became so available that fire destroyed huge swathes of trees. A particular type of fossil carbon called fusain is everywhere in Carboniferous fossil deposits, evidence of widespread forest fires.

But through their structure and chemistry, trees learned to survive and even to master fire, as well as the other elements, and out of the ashes of the Carboniferous forests a thousand new types of tree evolved. Trees had found a way of life and a timescale to operate on that proved immensely successful, and from their position as primary producers and earth's architects, even enormous herbivores like the dinosaurs could be tamed.

In 2018 I went back to Iraq with a clearance team to a field in Anbar province near the Euphrates River, where ISIS had laid a minefield of big improvised explosives. We started at 4 a.m. and by midday the sun was blisteringly hot, so I sheltered under a grove of date palms and talked to a man who, having lost two goats to mines, was understandably keen to make sure they were all safely cleared. Perhaps it was because I had already been jerked out of my comfort zone by the desert and by seeing the city of Ramadi flattened, the floors of the houses concertinaed on each other, but when he said that times were bad and that the palm trees were dying one by one, I felt I understood for the first time what environmental destruction was—the removal, piece by piece, tree by tree, of the chain of life. Everything—lack of water, dust storms, blazing heat, war—was acting together in concert to destroy these trees one by one, and without them everything else—soil, birds, plants—would go, too, and we would be left battling typhoons and dust storms, starving and thirsty, back in the world before trees, 385 million years ago.

What could possibly halt this? I realized the answer had been there four years earlier in Amedi, as we'd walked down the stone steps from the citadel, through the carved gate and into the woodland surrounding the town. The oaks I had looked at—multistemmed and gnarled, with leaves and trunks eerily different from and yet so similar to the oak woodlands of Devon over 1,800 miles away—had lived for a hundred, two hundred, maybe a thousand years. They had adapted and shaped the humans, animals, and plants around them; their roots were breaking rocks up into earth, sharing useful chemicals; their interlaced twigs sensitive to a thousand chemical signals; their leaves, exquisitely adapted to their environment, releasing water vapor that formed into clouds and precipitated rain. Besides communicating with the trees nearest to them, they were part of a meta-organism, a treescape that had evolved and spread over thousands of years all around the world. We know the bigger picture, that in the early days on this storm-swept planet, trees created places of greater safety, areas of deep-rooted stability—and that, interlaced together, they are actively generating that stability still.

One of the more specific superpowers of the oak trees found near Amedi is that as they get to a certain age they can change sex. Most of the young trees are male and produce pollen, while the older ones are female and produce acorns. This allows the oaks to feel their way forward into the future, with the huge old female trees being pollinated by multiple young males. The established oak tree, successful at surviving, provides lots of fatty acorns with energy for the next generation, an environment the young trees can grow into and DNA with a proven track record of success over many seasons, even as the wind and climate have gradually altered around them. The sapling males, on the other hand, have only recently grown up, and their pollen contains DNA and modifying proteins that have responded successfully to new diseases and sudden transformations in water or climate. The resulting offspring, then, should have the best of both worlds: an innate ability to harness the elements tempered by recent vicissitudes, honed by the ruthless hand of annihilation year after year after year.

It is too easy to think of trees as passive because we cannot see what they are doing. Victims, because we can take a chainsaw and clear a hillside of them. Does our aesthetic appreciation of trees incline us to preserve them? Have trees in some way, indeed, *cultivated* that aesthetic sense? The greatest wisdom is understanding that appreciation and conservation are two sides of the same coin. What we call the environmental benefits of trees—air cleaning, prevention of flooding, even sequestration of carbon dioxide—are side effects of trees' abilities to shape the environment. Thankfully, we evolved into a tree-shaped environment, and our interests are aligned. Our existence, just like theirs, depends on it.

The agency of trees gives us hope for the future—at least if we can rein in our good intentions and let trees know best. Tree planting programs are one of the major ways that tree diseases spread. Planting eucalyptus, most pyrogenic of all trees, will not capture carbon long-term, and neither will pine trees. Rewilding too can allow trees to grow up that are more likely to be short-termist and burn or self-destruct. The specter of many unsuccessful forestry projects must be a lesson that one size can never fit all and that dogma destroys nature. Like all environmentalism this must be a global story of adaptation and exchange. Trees can't be everywhere, but they don't have to be—their moderating effect can stetch far beyond them, connecting Russia to America, China to Chile.

But our story begins off the coast of Africa, in damp forest on the tiny island of La Gomera, with the wild smell of laurel in our nostrils . . .

CHAPTER 1

Trees shaping water

How trees brought water down and sent it back up

There is a sky behind the forest, there are seas unbounded, seething, waves made from the foam of dreams and churned by hands of light.

Nāzik al-Malā'ikah, *Revolt Against the Sun*
(translated by Emily Drumsta)

The cloud tasted of almonds. The tree I was looking at grew up gray-barked and straight for almost 100 feet until its trunk dissolved into the thick mist, and everything about it—slim emerald leaves, dead flaking twigs, mossy junctions between the leaves—was dripping with great drops of water that bent a thousand reflections into smooth curves. It was a tented world of cloud and moss and deep green. I knew, but couldn't see, that on the ridge to either side the laurel trees and the clouds continued, but if I walked just a little way down the mountain the sky would clear, there'd be a hot sun in a blue sky, I would see the sea, be surrounded by cactus and spurge, and feel the baking heat. That alternate reality seemed a million miles away. It was February and I was in the Canary Islands, in the cloud forests of La Gomera.*

Trees are cloud chasers, and cloud forests—known for their constant, lingering mists—don't exist just in the Canaries; they are in Brazil and Costa Rica, China and Borneo, Australia and the Philippines. Cloud forests tend to occur in places where the landscape

---

* The Canary Islands sit off the coast of Africa, bathed by the desert winds off the Sahara and the trade winds coming northeast down from Europe. The island of La Gomera is one of the smallest, a little volcano that rises to 5,000 feet above sea level.

gathers water from the air—typically mountains next to the sea. But trees don't affect water only in cloud forests; all 3 trillion trees across the world have an effect on rainfall and waterflows above and below them.

In a sense, trees developed into trees to gain power over water. During photosynthesis, trees use packets of solar energy to split water into hydrogen and oxygen and transfer the electrons onto carbon dioxide so it can start to make sugars.[1] This means they need large quantities of both air and water—mutually exclusive unless you can operate vertically. In the earliest stages of their evolution as trees in water-logged environments, upward growth raised green parts of the plant above the water and into the air where it could photosynthesize, while in dry areas vertical growth downward allowed access to deeper water tables.[2]

Having started successfully, trees continued to evolve a tightly engineered anatomy to chase this advantage.* Above the ground, trees are rainmakers, growing tall to interrupt air flow with their leaves and trunks and branches, emitting volatile organic compounds like scents and alcohols to seed clouds, and releasing water vapor out of their stomata to cycle a gentle, consistent flow of moisture from the air. Below ground, their roots collect and redistribute water, ushering water down to the water table, lowering or raising the level of the water table to ensure they have just the right amount for their roots to be on stable ground. And in between, the tree can control and use the water within itself. Just as humans can reach up to pluck an apple, crunch hard to eat it, and bend down to plant the core, so trees use all these three capabilities to direct water across the earth.

Cumulatively, the earth's trees sweepingly adjust global water flow. Trees of all 73,000 species are constantly making minute adjustments, but normally the resultant changes are subtle, deniable, and easy for humans to ignore, or, as in the Amazon rainforest, on a scale

---

* The anatomy trees evolved is more tightly engineered than a human body is engineered, because rather than moving to avoid drought and flood they must control their own environment in dynamic equilibrium.

too enormous to be easily comprehended. I had gone to La Gomera because the dramatic change from desert to cloud forest is heightened by the extreme lengths to which the trees have gone, and continue to go, to maintain their clouds. You can see the water pouring off their branches, smell the terpenes seeding the cloud, and in the tangle of dark-green leaf shapes above your head it is obvious that you are looking at cloud catchers, branches designed to scoop out the belly of a cloud. What you can't see is the effect of transpiration—water molecules sucked up by the tree's roots hustling minerals through the trunk, up to the furthest leaves 100 feet above, and then with a final puff of energy evaporating off and out into the air. You can, however, feel it in the cool under the trees as heat departs with the water molecules that are heading up to swell the clouds.

But does water enable the trees, or did the trees enable the water? A little bit of both, but trees are good at clinging on where they are given even a hint of water to work with. When the climate changes around them, trees tend to evolve, so that, where possible, they outflank the change by getting even better at shaping water. Most of the trees that grow in the *monteverde* forests, for example, the mostly evergreen forests that flourish in mountainous areas, are rare laurels, surviving forest that diversified out of the chaos after the last great extinction 66 million years ago when an asteroid hit the earth at Chicxulub in Mexico, causing darkness, chaos, a cloak of iridium over the earth, and the extinction of the (non-avian) dinosaurs.

Before the extinction, forests were open canopied and dominated by gymnosperms (literally, "naked seeds"), which grow tall and straight and stiff, live long, and alter the abiotic environment—water, air, earth, and fire—dramatically. For the first 300 million years of tree existence, these were the trees that were shaping the world, and they still grow on six out of seven continents and thrive in some of the most inhospitable parts of the world. They include the pines and the firs and the larch, the monkey-puzzle, the yew and the Wollemi pine, the giant redwoods and the *Podocarpus*, the towering alerce and the mighty kauri, and also some of the most endangered trees on the planet. The asteroid strike marked the beginning of the end for many of them, and, because the trees that replaced them

differ fundamentally, I am, dear reader, going to crave your indul-
gence as we make a quick diversion into tree history. Stay with me,
because it's only a few pages, and essential to understanding not just
why trees have had an effect on water, but why they look and grow
as they do and the effect that this has had on the world.

There is a split down the middle of the tree world, separating the
conifers and the broadleaved trees. With a few exceptions (including the broadleaved deciduous alder trees, lover of riverbanks, which
produce little cones that float downriver; the *Casuarina* of India, the
Philippines, and Australia with their tubular, weeping leaves like a
green plume of horsehair and their spiky green cones; and *Platycarya
strobilacea*, found in fossils in the London clay and now happily embedded in forests in China and Korea and Vietnam, walnut-like until you
see its bristling brown cones), all trees with cones are gymnosperms.

On the other side of the split are the angiosperms. Two hundred
million years ago, a flowering plant developed under the gymnosperm canopy that later benefitted from the revolutionary impact of
the Chicxulub asteroid and became the diverse and abundant group
of plants we see today as the flowering plants. Some of these angiosperms (literally "contained seeds") grew up to be trees, and they
are mostly broadleaved and often deciduous, losing their wide leaves
once a year in colder or darker seasons. The angiosperms include the
oaks and the ash, the *Parrotia* and the baobab, the eucalyptus and the
laurels, the palm trees and the rhododendrons. The angio-, or con-
tainer part of their name, denotes the fact that they have a carpel: a
fleshy, adapted leaf that folds around the ovaries, meaning that pollen has to penetrate through some of the plant before it can produce
seed. This was an extraordinarily powerful mechanism, because it
gave the female plant—which here simply means the plant that will
produce seeds and is therefore investing most nutrients in the potential offspring—the power of selection, a dogma-defying ability
to determine its offspring's DNA and therefore influence evolution.

The result was diversity and flexibility, chemical and physical. It's not just imagination that makes angiosperms look more youthful and less staid than gymnosperms. Giant sequoias and other gymnosperms often have burls, huge shoots waiting to spring up if they hit the ground, and it is supposed that these are an adaptation to the trees being knocked over by dinosaurs. By contrast, an angiosperm will root in a thousand places—even a 100-year-old beech tree can send up shoots from its trunk if it falls over, and becomes a phoenix tree. Genetically too, angiosperms tend to be more flexible, happily duplicating their DNA and experimenting with new chemical compounds. The ability to produce flowers and fruit, as well as shorter timescales of reproduction, meant that angiosperms shaped biotic factors—bacteria, fungi, plants, animals, and probably also humans—more than gymnosperms.

In the dark of the impact winter that followed the asteroid, the gymnosperms, those old evergreen trees that smothered the earth in the Carboniferous period, suffered.*[3] Their leaves, big enough when carbon dioxide levels were high and the earth peaceful, were suddenly too small, the veins bringing water and nutrients to the leaves too rudimentary for a dramatically changed climate, and their long lifespans a disadvantage in a world of chaos. They were supplanted by the angiosperms, which could cycle water quickly and drop their leaves when necessary, conserving energy until the next opportunity arose.

As a result of the rise of the angiosperms, tree productivity increased hugely. Whereas gymnosperms grew (and continue to grow) in elegantly spaced forests with one imposing species dominant, the new angiosperm rainforests were crowded and multilayered, fast moving, and packed with diverse species. The laurel forest that arose in a humid period 40 to 15 million years ago is just one way tree ecosystems coevolved to bring down water together.[4]

The period laurels developed into is broadly known as the Tertiary, and the trees that survived the droughts and subsequent glaciation are, like the spouses of a dead climate, sometimes known by the Dickensian term the Tertiary Relics.[†] About 15 million years ago,

---

* Gymnosperms are still struggling—37 percent are in danger of extinction.
† This term is technically no longer in use, but I think it's expressive so continue to use it throughout.

monteverde forests, the green, tropical or subtropical montane cloud forests, were widespread across Europe and North Africa, the vegetation of the Mediterranean before the ice age and subsequent onset of the hot Mediterranean summers. But as ice spread across the northern hemisphere and locked up water, the climate became drier and most of the laurel forest died off, leaving tiny reminders of itself around the shores of the Mediterranean: the fragrant bay laurel of Italy used in *osso bucco*, the warty *Zelkova* of the rain-catching White Mountains of Crete, and the maple-leaved liquidambars of humid coastal Lycia, perfume-bottles of the ancient world.

Close to the sea, trees have moisture in the air to work with, and in the Canary Islands you can see, with dramatic clarity, how the cloud forest adapted the climate to itself. On La Gomera, only 200 miles from the Sahara, and 3,300 feet above the sea, this fragment of the earth's cloud forest precipitates a world better known to high mountain dwellers: a world of thick mist, high rainfall, and cool, changeable weather. It's also an area of massive diversity and high carbon capture. On the Garajonay peak of La Gomera I was in a place where you could see and feel trees manipulating water.[5] I stood under a holly and looked down over the northern side of the island. There were mackereled clouds blowing toward me, and as they approached they seemed to rise up. Three minutes later I was drenched.

I had arrived in La Gomera by boat on the island's south side, where dragon-trees sit in formal gardens and houses bask in the heat. In the hotel there was a spectacular dragon's blood tree, *Dracaena draco*, a spiky, canopied tree with dramatic gray leaves like rapiers, sap as red as carmine, and fragrant white flowers.* That night I went and furtively lay underneath the tree, looking up at the stars through its strange fractal branches. Just like the laurels on the hill, it is a survivor from one of the oldest ecosystems extant today, the warm, wet

---

* Which is technically an overgrown lily.

woodlands of the Tertiary, but unlike the laurels it has adapted to survive on small amounts of water and to transpire as little as possible. Its relatives cling on across Africa, further south on the Cape Verde islands, and on the island of Socotra in Yemen, where they form ravishingly beautiful woods across island slopes in the blazing heat of the Indian Ocean. Its more distant relatives have survived in China, Vietnam, and Thailand, stretched thinly around the world by the movement of tectonic plates, and there are even two species in Central America.★ When Columbus set sail in 1492 from San Sebastian for his first Atlantic crossing, massively understating the distances to his timorous crew and shying at every crab, bird, and sign of life he came across, *Dracaena*'s spears had already been influencing both sides of the Atlantic for millions of years.

How do different trees adapt to their distinct niches and still shape water? The fastest route is by adaptation of leaves, and particularly changes in the stomata—microscopic pores within the leaves that allow water, gases, and other chemicals in and out. I had looked closely at the gray rapier leaves of *Dracaena* under a scanning electron microscope in a laboratory at Oxford.[6] You can see scurfy layers of wax flaking off the cuticle surface, a neat tilework of rectangular cells, and navette-shaped pores deeply embedded in the surface of the leaf. If the leaf's cuticle is a defensive, Byzantine wall of cells between a tree and the outside air with all its perils—bacteria and fungi and their spores waiting to loot leaves of sugar—then the pores are conduits that can be opened to allow air in. But the cuticle is also designed to keep water in, and under drought conditions pores can become a liability, allowing more loss of water than a plant can stand.

The pores or stomata (from the Greek for "mouth") are gateways for carbon, and because they are so fundamental to tree survival they can't evolve quickly—it would be as risky as our lungs suddenly changing structure. This means that stomata of *Dracaena* all over the globe look very similar, but the dragon's blood tree uses the wax that protected those mouths from drowning in the fine, damp days of

★ One spread across Mexico and Belize, Panama and Colombia by howler monkeys, and one confined to eastern Cuba.

the Miocene to prevent water loss now. And its blade-like bunches of leaves[7] use the same smooth waxiness in the winter months, as they whisk around in the wind, to capture any thin wisps of fog that blow over it, absorbing droplets of water along the blades and into the leaf bases and stems to act as a war chest for the summer.* It then oozes vibrant blood-sap over the water-swollen leaf bases, a resin full of ring compounds—garnet-red flavonoids, steroidal saponins, and antimicrobial polyphenols—that both seal in water and repel any scrounging animals desperate to drink. Restorative, coagulant, and antifungal, these compounds soon attracted the notice of humans. Who would not bleed a tree for compounds like these?

Not all the water dragon's blood trees scythe out of the air is stored. Much of it drops onto the soil, helping other plants to establish themselves. After high rainfall the navette-shaped stomata open, dramatically increasing the amount of water vapor they release through transpiration, and seeding the air for the next wave of cloud. Even with these adaptations they can struggle for survival. As humans chase them to still drier, higher, steeper places—in some areas literally off cliffs—their wild range is shrinking rapidly. In gardens, however, like in this one in La Gomera, they are flourishing, fed by water taken from springs and groundwater, with half their adaptations redundant.

Early the next morning, with dried fruit and water, I walked out of town, inland up the bare concrete-channeled river into the valley of the volcano. At first, little grew apart from Canary Island palms, *Phoenix canariensis*, lining the river, and I luxuriated like a lizard in the baking sun. Soon there were signs of irrigation; fruit trees and lush vines by the brooks and streams. Then, as I left the stream and turned right onto the zigzagging path that led steeply up the edge of the volcano itself, everything started to change. It was like walking

---

* Abrasion by sand and other wear and tear of the cuticle can help the water to settle and be engulfed.

into an Instagram filter of green. Mosses, hollies, everything turned from grass-green to emerald-green; the vegetation thickened, trees thickened and appeared in every crack in the rock, epiphytes started growing on them, leaves became spine-like or like slim dark ovals, branches and trunks started to occupy every available piece of space and, finally, as I walked over the ridge of the mountain and looked down at the sea on the other side of the island, I was in the cloud: deep, dripping shade.

Laurels were all around me, 100 feet tall with smooth brown trunks and spear-shaped leaves. They smelled aromatic and woody, almond-bitter. Another cloud came over us. I looked at the laurel nearest me, an *Ocotea foetens*, with smooth bark and branches supplicating the fog I could see settling on its leaves and coalescing and pearling on the bryophytes that grew all over it, until after a little while rain was dripping all around me, water flowing from the leaves of the trees like a fountain. I wasn't the first person to notice this; a tree of the same species called Garoé was a totem to the first peoples of El Hierro, the Bimbache.* A Spanish observer, marveling at the tree, wrote, "There is always a little cloud over the tree . . . all the leaves and branches drip, all night and day but more in the mornings and afternoons."[8]

It was like watching a tree gently milk a cloud. But beautiful as it was, I knew that the mechanical, tangible process of a tree physically condensing water in front of me like an alchemist with his still was unimportant compared to what was happening above my head, above the canopy of the trees. The invisible chemicals I could smell

---

* The Bimbache were relatives of the Berbers and were flourishing when the French explorer Jean de Bethencourt landed on the island in 1403. He described the trees that drip always with clean and beautiful water, collected by a trench next to the trees. One tree in particular—called Garoé—was sacred to the Bimbache and was considered the "fountain tree." It was not very tall, but it had long horizontal branches directed away from the trade winds, which caught the moisture in the air and allowed it to act as a water source all year round. It was a totem to the Bimbache, and in the first uneasy days of Spanish rule, Spaniards incorporated it into the Hierro coat of arms as a green tree on water with its head in the clouds pouring water from its leaves and called it the "saint-tree."

coming off the trees—those tantalizing hints of almond and camphor and cinnamon toying with my nose—were acting high up to
bring down the clouds that the trees could clutch at. Small, ringed
molecules of carbon were acting like grit in an oyster, seeding water
droplets and creating clouds. Just as for centuries humans have sent
incense and burnt offerings up to heaven as a sacrifice, so some of the
most finely wrought chemicals that trees make were evanescing into
the air, a sacrifice of energy to ensure reciprocal gains. The difference
is that when trees do it there is a scientifically measurable response.

In the tropics an estimated 30 to 50 percent of trees send up these
chemicals. In the Canary Island *monteverde* forests no one has done the
research yet, but my hunch is here the number is more like 80 percent. In the little ridge of forest I was among, not only the *Ocotea foetens* I was looking at but also the tree next to it, *Rhamnus glandulosa*,★
and the tree beyond that, *Apollonius barbujana* subsp. *ceballosi*, smelled
strongly, a sure sign they were sacrificing some interesting molecules.
The molecules trees use to cloud seed are called volatile organic compounds or VOCs: volatile, because they evaporate easily; organic,
because they're made of carbon; and compounds, because their
chemical makeup is a fascinating, tangled complex of many bonded
atoms. Trees have tweaked these chemicals for thousands of years,
but we still know few specifics about how they operate. All we know
is that they can have different effects.

Let's follow a phenylpropanoid molecule as it seeds a cloud. A
sacrifice of *Apollonius*, a tree related to avocado, 2-(3-methoxy-4-
hydroxyphenyl)-1,3-propanediol's structure was defined in 1995 and
(considering the name!) it looks simple: a hexagonal ring of carbon,
with one carbon arm sticking out, and various prickles of hydrogen and oxygen attached. It travels up and down the tree trunk surrounded by water, but when shoved into the free air of the stomata it
releases itself from the surrounding molecules and heads up into the
atmosphere. After minutes or hours, it will normally have reacted,
often with ozone or oxygen, sometimes with nitrogen oxides or a

---

★ *Rhamnus glandulosa* is called this because it has glands in the axils of the veins—the
stubs where they join onto the twigs of the tree.

human pollutant, sometimes with other compounds, to make a tiny solid or liquid particle. And then, inexorably, water molecules start to cluster to it, until like a small planet it plunges into a liquid state and falls as rain.

I could smell it, I was sure, that night, as I lay freezing in my bivvy bag on a flattish patch of moss under a Canary Island holly and soaked to the skin, I could certainly feel its effects. Pine and cedar, oak and laurel: those smells floating on the mist are one of the unappreciated superpowers of trees. Our noses are unable to detect all the hundreds to thousands of volatile organic compounds[9] trees produce that are a fundamental part of their immune systems, their communication systems, and their ability to influence water and pull down rain from the clouds. They are the thread running through all parts of this book, the mercurial quicksilver that runs through the veins of trees and allows the immobile solidity of a tree to act at a distance and cast a web up into the air or under the ground. Most of the laurels in the forest had been isolated on La Gomera for so long that they are distinct, very rare subspecies, but in this they are behaving like trees all over the world.

So far, few scientists have researched volatile organic compounds in the real world, so in 2014 the Amazon Tall Tower Observatory project set out to discover more about these chemicals that trees release. Brazilian scientists built a spindly tower reaching high above the canopy of the Amazon rainforest, and then captured air samples and measured concentrations of different organic compounds, their effects on the clouds, and how they interact with human pollution. Flying a small plane over the city of Manaus,[10] the only big city for hundreds of miles, researchers could detect the plume of anthropogenic pollution, and established that human pollution interacts with VOCs produced by trees to produce strange and dramatic results: deluges and floods. They started to define some of the staggering variations in the ways different trees release VOCs and found that almost everything had an effect: species, temperature, light, leaf age, and carbon dioxide concentration near the tree. Some trees were absorbing VOCs as well as releasing them. The research also revealed mechanisms of trees shaping water that had been overlooked by satellites.★ One massive discovery was that isoprene, the most common VOC released by all tree species, was three times more abundant than thought. The measurements also showed that trees released more isoprene under conditions of drought or stress, a crucial discovery because it suggested that trees were acting to encourage rain when there was a shortage, thereby stabilizing the ecosystem.[†]

𝕯

If trees can call down rain, does deforestation lead to drought? The Russian author Anton Chekhov thought so. In his short story *Pan Pipes*, an old shepherd bemoans the way that after the forests were cut down the rain dried up:

> "And what became of all them little streams?" he said. "In this very wood there used to be a stream with so much water in it the peasants only had to dip their creels in it to catch pike, and wild duck used to winter there. But even at spring flood there's no decent water in it now."

Most Victorian scientists also believed this to be the case. The drought and famine of 1877–9 in Bengal prompted a lot of work to investigate the links between deforestation and rainfall, and the Yanomami people of the Amazon and the Bishnoi of India are among many others who believe that drought comes if trees are removed.

Curiously, however, it has never been proved, and some have even suggested the opposite.[11] The world's forests are a collection of

---

★ The Amazon Tall Tower Observatory project discovered that the biggest mistakes found in models were the result of them ignoring the massive variations between tree species.

† Volatile organic compounds vary from tree to tree, environment to environment, and they are one of the ways that trees can speed up their own evolution. The gases can affect trees epigenetically, which means using proteins to amp up genes, and because they work in a modular way, doubling the genes for them can alter their size or the amount that is created. Most of all, they are very dependent on temperature and light, so they are often moderated by and moderators of circadian rhythms. They make good feedback loops.

cycles, confusing, stochastic, made up of the interactions of a million different organisms living on a complex, time-warped surface. There is no model widely accepted to show how forests promote rainfall. In the absence of such a theory it somehow became a given that rainfall decreased exponentially across continents, according to the prevalence not of vegetation but of mountains or lakes. Although the modeling did not fit actual rainfall data, the huge number of factors involved meant something—hills, lakes, temperature—could always be used to explain the numbers.

One idea that did fit that has consequently generated a lot of traction is the biotic pump theory: that trees use transpiration to cycle water inland from the coast—as long as tree cover remains unbroken. When trees first spread their leaves, they encountered a problem. Absorbing energy from the sun was necessary for photosynthesis, but the heat the sun generated could destroy the proteins in the leaf. Trees survived by releasing water from their leaves; as the water molecules evaporated, they took the sun's energy with them, allowing the leaf to cool down. This means that trees release a lot—almost 97 percent—of the water they take up just to cool themselves down. A single tree can give off 260 gallons of water a day, and as this water rises and cools it condenses, lowering the air pressure. If this happens next to the sea, moisture-laden air is sucked in over the forest, setting up a cycle that can continue for miles inland.

The theory was formulated during fieldwork carried out by Anastassia Makarieva, a theoretical physicist working in the Petersburg Nuclear Physics Institute. During her holidays she would travel to the north coast of Russia to spend time in the wet larch and pine forests near the Kara Sea, and she became intrigued by the physics of how quite so much rain and snow made its way south along the Yenisey River basin and inland as far as Mongolia. After more than sixty months of field research in the remote north, she formulated a theory that she published in 2007 with her colleague Viktor Gorshkov. Water cannot make it far inland in deforested continents, but rain penetrates far into river basins like the Yenisey and Amazon; "this points to the existence of an active biotic pump," they wrote, "transporting atmospheric moisture inland from the ocean."[12]

In other words, just as raising the handle on a village pump creates a half-vacuum and causes the water to rush up and fill the void, so trees, by seeding clouds above themselves through transpiration, were causing a reduction in pressure that sucked in water-laden air horizontally from over the ocean—air that in its turn would condense and fall as rain. By setting up a cycle of pressure gradients, a continuous forest from the coast to the center of a continent creates an airstream laden with water—a flying river.

These rivers are so strong over the Amazon that they can be seen in satellite pictures, which show clouds forming over parts of the rainforest like a white ticker tape of Morse code: rain and transpiration, rain and transpiration, allowing the rain to penetrate far inland.* The multiple layers of the rainforest canopy—the trees of the emergent layer such as *Dinizia excelsa*, which grows almost as tall as Big Ben, the canopy-layer trees like the rubber tree, *Hevea brasiliensis*, and the understory, filled with trees like *Theobroma cacao*, the chocolate tree—all produce turbulence in what is called "the clothes-line effect." The amount of rain evaporating off leaves increases, while transpiration adds to rising water vapor. Twenty billion tons of water vapor rise from the Amazon each day and fall again as rain, and gradually a continual line of rainforest can draw water far into a continental interior, allowing flat inland areas to be well forested. Experiments run in the context of Brazil have shown that if deforestation of just 1 percent happens by the coast, rainfall 125 miles inland will reduce by 0.01 inch *per month*.[13]

Can a few trees planted by humans really have an impact? When I was advising on a climate change documentary in Turkey, we went deep inland into Anatolia to film near Konya, the traditional home of the Whirling Dervishes. I went to talk about the environmental problems the region was facing with the head of the Dervish lodge. He decried defeatism and quoted Mohammed: "Even if you know that there will be an apocalypse tomorrow you should still plant a

---

* In 2009, Antonio Nobre, a Brazilian climate scientist, stated that without the flying river, much of southern Brazil, which produces approximately 70 percent of the country's gross national product, would be arid desert.

tree today." Just down the road, in this otherwise flat arid landscape, we met the farmers who have done just that—a consortium of sugar beet farmers who, with funding from Konya Şeker, one of Turkey's biggest sweet companies, have planted no fewer than 24 million trees on their land. Their new woods of walnut and plane, beech and cedar were truly impressive, but most of all the farmers claimed that rainfall had measurably increased as a result of their young forest. "Twenty-five years ago—we had hardly any rain," said one farmer. "But now our rainfall is increasing by about a third each year as the trees grow up."★

This opens up an intriguing possibility. In their first paper, Makarieva and Gorshkov used Australia as a possible example of human-driven drying that spread rapidly across a continent as a result of human deforestation by fire next to the coast. Another oft-quoted example of early deforestation is the Sumerian poem *The Epic of Gilgamesh*, which describes the titular hero cutting down the cedar of Lebanon forests that once covered all the high land around the eastern Mediterranean—the Jebel Liban in Lebanon, the Jabal an Nusariyah in Syria, and the Taurus in Turkey. The area around Konya is now a rolling landscape of bare rock and arid fields, but once upon a time the massed trunks and dark needles of cedar trees would have covered the whole coast of southwestern Turkey and released water vapor that would have precipitated rain and perhaps transferred water inland as far as Konya. If rain is already being attracted back as a result of the tree-planting efforts of Konya Şeker, is there a possibility that tree planting in Turkey, Lebanon, and Syria could alleviate the water shortages of the Middle East? Evidence suggests that coastal forests were already broken up by Neolithic times, and there is archaeological evidence from around the same period showing that climate change led to the abandonment of some settlements in eastern Syria. But it's impossible to know whether replacement of

---

★ Many companies claim that they can already see the advantages of having planted thousands of trees. Brazil's state water company, Sabesp, for example, has planted 174 square miles of trees on its land and claims to have already seen the benefits of an increase in rainfall.

these forests would affect how dry Iraq is today, just as it's hard to know whether Australia would still have dried out naturally had it remained untouched by humans and fire. What we do know is that trees, by their very identity, fight to maintain the level of water they enjoy.

$$\mathcal{B}$$

How did trees start to shape water? And why did this happen? Imagine a world before trees. Three hundred and ninety million years ago, when trees first developed on land, there was no grass, no soil, no rivers, and certainly no animals. High mountains reared up and crashed down, storms swept in from the sea, and rain on young rock fled in torrents to the sea.*

The nearest picture we can get today to such chaos is in newly forming mountain ranges, where raw rock is pushed up onto the surface. I spent time in the Hindu Kush in the Chitral province of northern Pakistan, hunting for a rare bulb, *Fritillaria chitralensis*, and was amazed by how tense this young mountain range felt, quivering on the verge of collapse as it was pushed upward. Young mountains, in their hard, brittle chaos, are some of the scariest places on earth. The rivers flood frequently with the sediment of intense landslides, and I woke up like a shot every night for a week, palpitating to the tremor of earthquakes.

By contrast, the Catskill Mountains feel sleepy and peaceful. *Time is a great healer*, I found myself thinking as I admired the red of the maple trees and osiers out of the window as the train wound up the Hudson Valley from New York, contrasting in my mind the rounded outlines of the ancient Catskills with the spikes of the still-growing Hindu Kush. But it's not just time that smooths over the wounds of mountains—it's also life. The Hudson River drifts gently past banks bound together with osiers and hickory, and water from the weath-

---

* There is a sketch by Leonardo da Vinci showing a torrent like this—a torrent with zero resistance.

ered Catskills seeps slowly down from mountains thickly clothed with beech and hemlock trees.

This peaceful scene is the culmination of trees pacifying the landscape from the earliest stages of their development, a process that we know about from proofs found in the rock of the Catskills. Traced on the rock are the fossilized root systems and stumps of some of the oldest fossil forests in the world. These are trees that grew 390 million years ago, at a time when Europe was grinding hard against America. At the constructive plate boundary, steep, volatile mountains as high as the Andes and as tense as the Hindu Kush were springing up; mountains that split under frequent lightning strikes and earthquakes to send landslides westward into a shallow sea.

Evidence of this vanished world traced on Devonian rocks tells us that all was chaos; in some places there was a hard crust on the earth, made up of bacteria and fungi, but it was often swept away. Rivers rarely meandered: they raced downhill as fast as possible, leaving a sediment pattern like a backward fish scale. Slower rivers moved in a braided pattern, soaking all available land, parting and rejoining without a clear channel. The right amount of water was essential to the life of the earliest trees: too much, and they were unable to stand up; too little, and they could not photosynthesize. Torrents toppled or snapped them, so the very early trees started to establish themselves in the most peaceful place they could find: the shallows at the edge of the sea. There they grew into tentative forests.[14]

The fossils of these earliest forests are kept and studied in the New York State Museum in Albany, a vast concrete spaceship built by Governor Rockefeller that landed incongruously in the 1970s on the beautiful city where Henry James grew up.* I went there to visit Bill Stein and the team who look after and research the trees and was whisked through ten locked doors into what felt like a bunker, full of huge tree stumps like giant elephant's feet and exquisitely preserved slabs of fossil tree. Behind a cabinet lay the stone imprint of an outsize clubmoss, 16 feet long and with strange bark like snakeskin.

---

* It is also where one of my favorite Henry James characters, Isabel Archer, started to "affront her destiny" at the beginning of *The Portrait of a Lady*.

A fern-like sheaf of leaves growing in the gravity-defying fan of a Mohican was impressed on a rock. Most impressive of all, on the wall was an aerial picture of a huge beige-red fossil-root footprint, made 385 million years ago by *Archaeopteris,* the tree that existed 150 million years before the dinosaurs. Amazingly, this footprint is a vast outline of crystallized organic matter set in green siltstone of the first true root-network known, 36 feet wide and sometimes diving 5 feet into the earth. *Archaeopteris* was the aunt of all the trees alive today: a ginkgo-leaved beauty with a feathery umbrella of fishtail leaves that grew to 78 feet tall, and appears to be the first group of plants to produce roots that, as with the trees we see around us, were reinforced and maintained *and directed* over multiple seasons, as well as sending out ephemeral little feeder roots into the soil.

These roots grew out year on year into a complete network worthy of the tree of souls in *Avatar;* main roots and subroots and minuscule roots stretching in a circle, just as the tree above the ground grew up into a eustelic,* long-lived, and wide-reaching tree.

Such trees apparently needed to develop a wide reach, sucking up more water than their earlier cousins. In a circular process typical of modern trees, the greater reach and depth of the roots allowed them to lower the water table and dry out the area in which they lived, making it more stable and less susceptible to flooding. The roots stabilized the soil and canalized water edges, just as nowadays we might put down railroad ties to stop a stream breaking its banks. And as these trees spread, growing taller and becoming the dominant species in this ecosystem over the next 20 million years, the striking consequence was the shaping of rivers, to encourage them, over the course of the Devonian period, to meander more and more, establishing muddy floodplains and fertile stable areas next to the coast. By 298 million years ago, tree roots had stabilized sediment

---

* The technical term for a tree with well-defined veins and trunk.

so much that relatively narrow and defined channels were the norm, and trees could grow tall, along their firm banks.

This is the footprint that was excavated at the Cairo quarry in the Catskills.* Unlike some of the other fossil forest sites uncovered nearby, which were flooded repeatedly, the *Archaeopteris* at Cairo was buried by just one serious inundation, possibly caused by a small tidal wave. A little fish brought in by the water was trapped by the roots and fossilized there, and the same wave seems to have shattered and killed the majority of trees at the site.[15] As well as *Archaeopteris* there are traces of trees called *Eospermatopteris*, tree-fern-like with elephant-feet stumps that sat in swampy ground, putting out simple roots that emerged and died, never growing further out than 3 feet. There were also some mystery trees, possibly lycopsids, species that looked like a cross between a giant clubmoss and a horsetail.

Unlike some other fossil forest sites in the area, there is evidence of the forest site at Cairo being relatively dry at one side, and the roots of *Archaeopteris* seem to grow toward the water.[16] Their ability to move water across the ground through their root system was the equivalent of them being able to pipe water from place to place. It was a massive evolutionary advantage, and the increase in biomass it prompted on land made *Archaeopteris* the fairy godmother of fresh-water fish, which started to develop during this period, nourished by the leaf litter from its branches and breathing in the more reliable flow of rivers.[†]

The 36-foot reach of *Archaeopteris* roots marked the start of trees hydro-engineering on a huge scale, but this was only the thin end of the wedge. Bill Stein's careful study of the earliest tree-like fossils he

---

* This imprint was found by Charles Ver Straeten and Frank Mannolini in a quarry near Cairo, New York.

† There were still no herbivores, but detritovores started to develop in the stabilized earth, and *Archaeopteris* was possibly the architect of the first mass extinction caused by trees changing the air concentration.

unearthed convinced him that in order for the energetics of trees to work, they needed to develop a leaf-like structure early on, with the fleshy photosynthetic cells of the parenchyma stretching across the veins. Suddenly the surface area over which light could be absorbed and water released was many times what it had been; to counteract the heat of the captured sunlight, dramatically increased levels of transpiration were necessary. Over the next 300 million years, as trees developed leaves with greater surface areas, they also developed more effective roots and shoots to ensure that water and nutrients got to the leaf, even beyond the technical barometric limit of 30 feet.[17] Sunlight was captured, sugar was made, and optimal conditions for growth were maintained. In all of this, water was necessary for trees, and as the ways they obtained it became more sophisticated, so they shaped more and more of the channels, above and below ground, that water ran along. Some trees even started to link up underground.*

Trees are known to detect or "hear" water underground; they will often move toward (and into) pipes that contain it.[18] Many tree species reach down to the water table—the layer of water that sits under all the earth at approximately the level water would sit in a well—to find more water than they can absorb from the surrounding soil.† Even in the UK this is proving to be more common than most people realize. A stand of Scots pine in the Highlands, for example, will certainly take up water from the water table in dry summers and possibly also at other times of year.[19] Eucalyptus trees can lower the water table at a rate of 2 feet or more in dry summers and were often planted specifically for that reason,[20] but eucalyptus is unusual in that it will lower the water table by 33 feet over ten years and not restore it. Mostly, trees will refill the water table by channeling water down into it by infiltration; rain can run unhindered down their trunks and then continue through the channels made by their roots deep into the ground and into the water table. Even trees without deep taproots will channel water down away from the surface, just as the first

---

* The great example of this is the *Lepidodendron*, which meshed its roots into a matrix in order to sit in swamps.
† In Tanzania people calculate the taproot of trees—normally three-quarters of their height—and prepare to dig their wells that deep.

Archaeopteris roots did, as it prevents the soil becoming waterlogged and stops flash flooding.

When I was working in Iraq, I remember the joy of coming across single examples of the Euphrates poplar, *Populus euphratica*, by the side of the road—a trunk with olive-green bark and a fluttering broad canopy with a dramatic sheep-grazing line, red leaved in the autumn and silver leaved in the spring. These leaves twisted and turned in the wind, broad and deeply toothed. The tree would grow isolated on mountain sides, far from any water, but it also grew along the Euphrates and the Tigris, and here the poplar would sucker, the root traveling along the water, stabilizing the riverbank and reaching deep down into the groundwater until it sent up a new shoot, while its leaves changed to look much more like those of a willow. It's a successful survival technique in areas with extremes of water. While the rarely suckering Syrian ash, *Fraxinus syriaca*, has all but disappeared from Iraqi riversides in the last fifty years, partly in response to a massive drop in water levels in the Tigris and Euphrates, the poplar is flourishing.★

The Euphrates poplar seemed remarkable enough there, in otherwise friable and cracked, baked earth, but years later I saw pictures of the *Populus euphratica* of the Taklamakan Desert—the same species growing out of bare sand dunes, a whole row of them along an invisible river. The poplars of the Taklamakan reach down feet into the water just above the water table—the *phreatic layer*, originally named after the Greek word for spring—and when one tree hits water they will share it with the other trees they are attached to: their clones, which can spread across as much as 300 acres.†‡ Pando,

---

★ Near Ur, on the Tigris, an Iraqi charity is planting tamarisk trees to reverse salination.

† There are few trees that have the ability to reach down deep enough to hit the water in these ever-shifting sands. Others include the four species of the genus *Alhagi* (literally "pilgrim" in Arabic)—spiny, defensive trees that are spread across desert landscapes from the Taklamakan to Greece and down into North Africa as far as Chad.

‡ In deserts, plants not only fight to get water, they also fight to resist the onslaught of animals that want the water themselves, and so shaping water becomes an active

the star example of this ability, is a male aspen, *Populus tremuloides*, that, by suckering, has spread across 106 acres of Utah and weighs 6,000 tons altogether—the heaviest known organism and one of the oldest: perhaps as much as a million years old.[*] During this time, Pando will have sent out again and again a lateral root to run underground until it reached a fresh area of moisture, light, and soil. As the old trunk died it would have sent up an identical trunk, using the energy generated by the rest of the tree. The whole process allows trees to endure drought and stabilize sand or endure water and stabilize dry land. It seems, in fact, a miracle solution to engineering water below ground; a way of spreading your bets.

Another way of spreading your bets is to release millions of seeds, like a willow, but for aspens creating a new clone is not particularly easy.[21] The question of when a seedling can be laid down is largely dictated by water. In the Taklamakan, for example, seedlings can only germinate during rare flooding events, and then the root must grow like smoke to keep up with the floodwater as it sinks back to the water table again.[†] The effort is worth it: because some trees can reach down 65 feet (the height of a four-story building) to suck up the water they need and share it with others that may be better set up for photosynthesis, some poplar monocultures in the middle of the Taklamakan Desert are more productive than temperate forest.[22]

As equalizers of water over a large area, suckering trees make up the majority of species growing in areas with extremes of water—the very dry and the very wet. Many of the world's rivers are lined by suckering trees, and so often has it been an evolutionary advantage

---

part of their survival, often requiring hundreds of years of patience. It's a strategy that works in the wet as well as the dry; in Xinjiang, far from growing out of the desert, a forest of 4,363 square miles grows out of the floodplain of the Tarim River.

[*] This would require Pando to have survived multiple glaciations, so most estimates stand at around 10,000 years old.

[†] While in Scotland, coordinated flowering events of aspens happen very rarely—recently these events have only occurred in 1996 and 2019 as they rely on very hot sunny summers. In a similar way, poplars in Utah have rarely established new colonies after the Ice Age lowered temperatures 12,000 years ago.

that it's responsible for many of the extraordinary abilities we think of as particularly tree-ish: the ability to grow from cuttings, grafting, and budding.

The first suckering tree developed in Antarctica 245 million years ago in the Triassic,[23] a now extinct conifer called *Notophytum krauselii*. The extraordinary fossils of this extraordinary tree, which had slim, fleshy green leaves not unlike the laurels of the Canaries and grew a little like the present-day southern hemisphere family of the Podocarpaceae, were found in what is now one of the most lifeless parts of the world, an Antarctic mountain called the Fremouw Peak. It is bare even of snow and of most living things except a few lichens and mosses, but it is crammed full of dead animals and plants: a perfectly fossilized record of a lush riverine community full of reptiles and conifers, ferns, horsetails, and cycads.

While all this was happening, the rest of the gymnosperms were growing a bit like giant sequoias nowadays: slowly, well spaced, structurally. The occurrence of suckering here, long before it appeared in trees across the rest of the world, tells us something about the reasons it developed. Was there something happening in the warm forests of the south, possibly as a result of this riverine environment, in which frequent floods made it important for a tree destroyed in one place to shoot up again from its roots further along the riverbank? It's possible that the need to take advantage of the long days of summer, so far away from the equator, made it advantageous for a destroyed tree to be able to resprout quickly. But being a clone is a dangerous business, and *Notophytum* became extinct; perhaps its ability to distinguish between above and below ground wasn't good enough. However, the ability to sucker evolved again. When angiosperms, the flowering plants, started to take over from their coniferous cousins, suddenly suckering was everywhere.[24]

The pinnacle of tree water-engineering comes with the mangroves, which grow tall to balance the need for water with the need for air, but have also adapted their roots to spread sideways to stabilize their

environment, and in some cases to obtain air for cellular respiration.[25] As a result only specially adapted trees can cope with being
constantly submerged and must find a way to get air to their roots,
usually by using the gray, periscope-like pneumatophores that poke
out of the water all around the photosynthesizing tree. Mangroves
are the standout performers in the chaos of the intertidal zone, tol-
erating both waves and salt, and occupying more than 54,000 square
miles worldwide. In the warmer world of the Eocene they were
everywhere—*Nypa* mangroves are found in fossils in the London
clay—but they require a lot of energy to manipulate water as thoroughly as they do, and so as the world cooled their distribution slunk
back to the tropics.

From a small boat you can see why mangrove forests are so popular in the world of carbon credits. They look like a swirling mass of
straws, sucking up water and sediment through their roots, while also
sucking down air through the pneumatophores, making it silently
into tree-mass. In so doing, their roots act as a skeleton around which
sediment will agglutinate, absorbing the energy of the water, and the
trees push back the sea. Few mangrove forests are as powerful as the
Ilha Grande mangroves, which in a densely populated area of Brazil
provide essential protection from the Atlantic swell to many small
islands and a fragile coast. The mangroves of Ilha Grande are mainly
red mangroves, *Rhizophora mangle*, but white mangroves, *Laguncularia
racemosa*, and black mangroves, *Avicennia schaueriana*, also grow there,
working in series to stabilize the coast. Red mangroves survive in the
most inundated areas, propping themselves up on hoop-legs and absorbing air through lenticels—lung-like gaps in their bark that contain porous, spongy cells. They store the air directly in their roots for
when they are inundated, and waterproof the roots with suberin, a
sort of wax, which also allows the mangrove to ultrafiltrate salt out
of the water it takes up.*

Not only do mangroves help protect coastal cities against rising sea
levels, but they can also be the first step in the transfer of water in-

---

* A rubbery strip called the Casparian strip forces water to move through a sort of
selection filter, allowing the tree to filter out the salt.

land. My father used to tell a story of flying a small plane from Mombasa to Lamu through a clear sky, and how the plane, which had held steady over sand and sea, started to rock furiously as it passed over the heated updraught off the black coastal band of the mangrove trees. These convection currents and the tropical rainstorms they produce allow mangroves to exist on minimal fresh water.[26]

In the form of mangroves, trees have returned to colonize the water they left 400 million years ago, having evolved upward and downward, but most of all chemically, beyond all conception of the earliest trees. They have used their own false starts and dead ends of evolution, their own branching and diversity, to revolutionize the limits of chemical and physical manipulation of water. But from the *Archaeopteris* trees of Cairo onward, trees have defied gravity to spread water thinly and in a form they can use across the globe.

CHAPTER 2

Trees shaping soil

How trees invented earth in their search for growth

When the world was still all new, when the sky was fresh and the earth not yet soiled, when trees breathed through the centuries . . .

Intizar Husain, *Basti* (translated by Frances Pritchett)

Give a tree a lump of rock and some time, and it will make its own habitat. Trees will break rock, digest it to extract phosphorus and other minerals, gradually beat in carbon and nitrogen from the air, enervate and stabilize the resultant soil, and mold themselves a home of deep-rooted stability. Botanists call the world under trees the rhizosphere, literally the world of the root, and it remains an alien world of unexpected twists and turns, full of enigmas as well as intriguing answers.

When trees started to develop 400 million years ago, mats of fungi and bacteria formed a crust on the rock of the earth, but it was mainly abiotic factors—ice and water—that split rock or ground it into sand and dust. Why did tree roots start to grow down? There are various explanations, including a search for the water table and the need to anchor a photosynthesizing tree above the ground. But the most fundamental is that a plant's survival entails growth. Unlike animals, which can dismantle an old cell and replace it with a new cell in the same place, plants must keep their cellular structures stiff like the bricks in a wall, and are confined to laying down a new cell next to the old one.[1] This is the reason that old trees often have a massive but hollow trunk, in which their own roots are digesting their own rotten heartwood. They can't replace tissue, only break it down to start again. In the case of a newly growing tree, root survival involves expanding out and down and breaking up anything in the way.

More fundamentally, why did trees evolve to grow into the rock and turn it to dust?[2] As tree roots start to grow and digest the rock that they move through, suddenly minerals are available to them that are essential for their growth (notably phosphate), for DNA, for cell membranes, and for moving energy around the plant easily. A feedback loop ensues, in which the top of the tree supplies the bottom with energy from the sun, and the bottom uses it to grind through the rock and supply the top with phosphate, which in turn helps the tree grow. Root hairs also provide a vast surface area for the osmosis of other mineral salts. Iron, zinc, magnesium, copper, cobalt, and potassium all cross from rock into the organic realm in this way.

Having broken up the rock, why do trees then bind up the rock again with carbon chains and make what we now call soil? Why did sediment start to build up on land rather than consistently sinking to the bottom of the sea? Why do trees recruit and house bacteria in specially adapted organs, feeding them until they enrich the soil around them with added nitrogen? And why does all this help to trap water and keep it evenly spread out above the water table? Why, in short, do trees shape the earth into something we would think of as habitable?

"I can show you fear in a handful of dust," wrote T. S. Eliot.[3] For humans, the difference between soil and dust is the difference between life and death. From one we can extract nutrition, essential energy; from the other we starve. Picture the *kemet*—the black soil of Egypt made fertile by the flooding of the Nile—as opposed to the *desheret*—the sterile red dust of the desert. It's not just that dust has no moisture, but that it cannot trap water without the complicated chains of carbon plants have pieced together using the energy of the sun as a binding agent. Whereas desert soil may have less than 0.5 percent soil carbon and almost no nitrogen, most very fertile soils like the *kemet* will have over 20 percent carbon and 0.01 percent nitrogen. Without energy from the sun packaged by plants into their organic matter, humans literally cannot keep the carbon chains of their lives together. Entropy, the pull of the universe's disorder, acts on us and we fall apart.

Trees work with dust and make it soil by the addition of carbon and nitrogen, thereby acting like a sort of ground bass, an enduring rhythm that gives life to countless other organisms. Much of the organic matter in the *kemet* comes from the trees at the Nile's source: the tropical forests of Uganda, the junipers of holy Gish Abay in Ethiopia, and the Ituri rainforest of the Democratic Republic of Congo. This organic matter is partly dead leaves, bark, and wood dropped by trees, but trees also actively add carbon to the soil in the form of root exudates.* When the *desheret*, laden with phosphate, is blown to treed places such as Crete, it is soon transformed into fertility. Trees are capable of weaving soil out of the dust, and although they cannot live without any water, they can create texture to trap it. From rock they can beat out a framework in which to work, by using only what they trap from the air, the same greenhouse gas that is spinning us along in the maelstrom of climate change.

Without this dynamic we would lack many of the landscape features we take for granted. Hills and meandering rivers and fertile floodplains seem inevitable, but they were created for the first time in what paleobotanists have christened the Devonian landscape factory. In a 2021 paper, Neil Davies and his coauthors analyzed a set of superbly preserved rocks in Svalbard and pieced together a picture of 380 million years ago: a Willy Wonka landscape of hills and hummocks and valleys, mud and sediment islands, strange constructions and peculiar patterns.[4] Logjams and tree roots were altering the landscape dramatically, the authors point out, when beavers were just a twinkle in evolution's eye.[†]

The forerunner of woody sediment, the organic matter that drops into the Nile and makes the *kemet* rich and black, makes its first appearance in these Svalbard rocks. So does muddy foreshore,

---

* Just as the leaf litter of *Archaeopteris* enabled the development of freshwater fish, so the trees of the Nile enabled Egyptian civilization.

† Specifically, logjams were altering the landscape about 385 million years ago. The common ancestor of the American and European beaver appears to have evolved around 10 million years ago.

and so does the first driftwood, blown gently against the seashore. The Oompa Loompas in this scenario are the trees: chaotic and fast growing, working out strange rules of dress—bark textures and leaf shapes, different configurations of rooting and branching. At around 20 feet tall they were a perfectly good size, industriously constructing new and extraordinary chemicals, and spreading swiftly across their new world. When they toppled over, their disrupted root plate left holes in the earth, and where their trunks landed they acted as bulwarks against the shifting sand. Their branches and leaves fell off quickly and often, and where they fell they baffled sediment, leading to different textures of earth and new ecological niches. The children in the factory trying the exciting new treats on offer were some of the first detritovores—millipedes and arthropods and the occasional worm, whose fossilized burrows work around the trees in lonely evocations of what was to come. Amazingly, the first trees lived in a world without herbivores. *Nothing had yet evolved to eat them.*

When I went to Albany in October 2023 to see the fossil forests found in Gilboa and Cairo, I had no idea that the record for the "oldest fossil forest" was about to be broken by a place forty minutes from where I grew up, a cliff I had biked to and scrambled over a hundred times in my childhood. The north coast of Devon is a place of dramatic cliffs and sudden chasms; gray stone is folded up into hairpin bends in the rock, small streams roar down from the sponge of Exmoor's heather. Hawthorn and oak woods cut deep channels in the rock, and the south coast of Wales in the distance always seems to steal the sunlight. The rocks on this coastline gave their name to the Devonian period, but after a few fossils had been collected and classified paleontologists moved on, distracted by the dramatic ichthyosaurs and dinosaurs of the Jurassic south coast. However, when a team from the Universities of Cambridge and Cardiff reassessed the fossils they found something interesting: the remains of a fossil forest from 390 million years ago—the oldest in the world.

I went to see them: imprints in the gray cliff, slim, moss-like trunks, about the same width as my wrist, with a snakeskin texture. Compared with those at Cairo and Gilboa, the trees were tiny—little sprigs shaped like palm trees, with twig-like leaves—but there were lots of them; called *Calamophyton,* in places as many as seventeen trunks were lying on top of one another. And these rocks are evidence of dramatic water engineering by the trees: the rocks held the traces of logjams that had trapped the sediment, and leafless trees had dropped twigs in profusion, which had slowed to a trickle the water flow underneath them.

$$\mathcal{D}$$

From the Devonian period onward, one of the major drivers of change in the global landscape were roots. What made trees reach out and down rather than remaining, like lichens or bacterial films, a layer on the rock of the earth? The earliest plants on dry land 445 million years ago sent out little extensions of themselves, rhizoids, to cling to the rock. Rhizoids are a bit like root hairs, delicate probes one cell thick that can suck up water and nutrients. Lift a clump of moss from the cracks of a pavement and you'll see them; they are still the organs mosses use to cling on to soil and rock and trees. As we have seen, larger vascular plants, particularly trees, needed a structure that could supply more water and act as an anchor. They developed a thicker, stronger, specialized root that branched and explored the world, pushed this way and that by chemicals, plant hormones called auxins. These spread up and down the root in response to water, light, pressure, and nutrients, triggering growth, cell division, a branch in the root.

This required a surer, quicker mechanism for detecting the center of the earth, the direction in which water and nutrients and stability were likely to be found. The mechanism that evolved is a microscopic equivalent of a ring of neolithic standing stones: statoliths, also known as amyloplasts, essentially heavy sacs of starch wrapped up in a membrane. These sink through cellular fluid to lie heavily along the bottom of the cell and are only found in certain cells at the root tip. Like the standing stones at Stonehenge, therefore, their presence indicates both the specialized place and the action to be performed. Imagine a henge, standing on a hilltop and visible for miles around. It attracts druids, who gather below it, and directs other people to the site. In the root tip, statolith piles at the bottom of the cell attract PIN-transport proteins, which shuffle through the membrane like druids striding across the surface of the earth to gather under the statoliths. Once gathered they usher plant hormones, auxins, across the membrane and into the cell. The auxins stop that side of the cell from elongating and the root bows to gravity and heads down.

This was a remarkable evolutionary advance, cementing plants' obsession with the earth and allowing them to reach up for the heavens. It allowed for delicate and efficient roots that would grow straight down where possible, or, if forced to, along an impenetrable rock while constantly probing it for any weakness. Almost any seed plants—whether rose or cress or oak tree—will use this mechanism, but gymnosperms developed a particularly strong pull earthward. Extremely tall trees like the Norway spruce and the American loblolly pine have two varieties of the PIN transport protein, caused by a gene duplication long ago, which means their roots sink further, and have, if you like, a stronger tendency to journey to the center of the earth.[5] This demonstrates the importance of trees being grounded. Norway spruce can grow to be 180 feet tall, but their growth toward the sun is impossible without a balancing predilection for the center of the earth.

Many trees have the ability to grow deep. The deepest recorded taproot is that of a wild fig, which grows down nearly 400 feet, 20 feet further than the tallest known tree, following gravity through rock and air in the echo caves of the Mpumalanga region in South Africa. The taproot is normally the first root to spring from a tree seed such as an acorn, growing straight down in an initial bid for survival to provide stability and water.* Later, most trees develop heart roots and lateral roots, which level off at a calculated point and quest into myriad networks of the soil, searching for a considerable amount of nitrogen and phosphorus, slightly less calcium, magnesium, and potassium, and trace amounts of aluminum and selenium, boron and sulfur, copper and zinc, iron and manganese and molybdenum. While extracting all these elements, they can't see what is

---

* Some trees never have a taproot—it depends on local conditions and tree species.

beyond their immediate environment, but they can sense concentration gradients. Trees below ground are blind explorers, but like all explorers they irrevocably alter the world as they move through it and sometimes they invert the world; the metals they extract can find themselves in the canopy 200 feet up. It's hard to picture the extent to which roots track down minerals without intent; in fact, their search can seem so directed that people have suggested that they can act as an integrating factor for trees, almost a brain. In the last paragraph of *The Power of Movements in Plants*, Charles Darwin and his son Francis write: "We believe that there is no structure in plants more wonderful, as far as its functions are concerned, than the tip of the radicle. If the tip be lightly pressed or burnt or cut, it transmits an influence to the upper adjoining part, causing it to bend away from the affected side; and, what is more surprising, the tip can distinguish between a slightly harder and softer object, by which it is simultaneously pressed on opposite sides." These mechanisms have now been well characterized at the minutest level—but the Darwins went on to say: "the tip of the radicle thus endowed, and having the power of directing the movements of the adjoining parts, acts like the brain of one of the lower animals; the brain being seated within the anterior end of the body; receiving impressions from the sense organs, and directing the several movements." "Our book on the movements of Plants will, I think, contain a good deal of new matter, but will be intolerably dull," wrote a morose Darwin to Joseph Hooker, director of Kew Gardens in 1879. Further research is making the Darwins' root-brain hypothesis, as it's called, more and more plausible, and the book no longer reads as intolerably dull. In 2024 it was proven that some non-neural cells in worms can remember routes to juicy liver that they learned when they were still attached to a brain, and then reteach the routes to a new brain, storing the information in electrical patterns within the cell. The idea of a multipronged brain, disseminated through the soil, is not so foolish when sparks of evidence seem to come together to suggest it: cells flaking off from the root cap moving through the soil exude carbon, sequester poisonous ions, and influence bacteria and fungi; roots can grow toward the sound, or at any rate the vibration of water; electrical impulses can shape biochemical responses faster than can be imagined. On the other hand, to ask whether roots have senses is in one way totally missing the point. It is for animals to sense and respond and hunt; trees have evolved to stand still, so they need merely exist and find and create. Bacteria and fungi can act as scouts, as good indicators of these chemical deposits, but they are tricksters with their own agenda—not to be trusted and liable to sequester nutrients for their own ends. Fundamentally, the roots must grow by themselves, their lust for depth moderated by the pattern of chemicals below the soil and their specific needs and not deterred by the hardest rocks.

Phosphorus is one of the elements essential for all life on earth, and it makes up a quarter of the nutrients a tree cannot make itself but has to find from rocks. It is necessary not just as a component in DNA and RNA, cell membranes and protein, but is also indispensable as an enabler of chemical energy transport—vital for the tree to transfer usable hits of energy to its growing root and shoot tips. In fact, much that trees do to shape soil is a side-effect of their hunger for phosphorus.

Phosphorus itself is a highly reactive element, but having reacted with four oxygen molecules it forms a stable ion, phosphate, which is found in small amounts across most of the rocks on earth, mainly in the form of the mineral apatite. Some of the most aggressive movements of trees on young soils, their most egregious rock crunching, can be linked to trees searching for inorganic phosphate, the raw phosphate in rocks, whereas some of the fiercest competition in old soils can be linked to the fight to recycle organic phosphate, the phosphate already linked to proteins or membranes or DNA.

In the hunt for phosphate, trees have been victims of their own success. The early sea was full of dissolved apatite, which is why so many chemicals central to life are based on it. But when trees started to spread across earth's surface during the Devonian period, 358 million years ago, they dramatically increased the rate of weathering of the earth's surface, and one of the unintended consequences of this was that phosphate became harder to find. Trees can secrete small carbon chains that work as acids to liberate phosphate from rock, or they can produce phosphatases, great lumpen proteins like nutcrackers that extract organic phosphate from plant remains. Either way, they must continually seek out phosphorus and try to lessen the leaching effect of water that sucks it out into the ocean.

Trees must work particularly hard to find phosphate on limestone soils, because phosphate and calcium tend to precipitate into a solid, preventing trees from taking up the phosphate in liquid form.* I write this sitting on a scree'd mountainside looking down on the circular sweep of the Omalos plain in Crete. Above me are the leaves of ambelitsiá or *Zelkova abelicea*, just turning a beautiful rust-red as the frost begins to bite, and in this environment they are the phosphate prospectors, the first trees getting to grips with the rock. The advantage for them is that there's little competition from other trees; the disadvantage that getting phosphate out of limestone is hard. They dissolve rock not only with organic acids but also with the mechanical action of their roots. The leaves are small and stiff and thick with a midrib and seven side veins on either side. Something has nibbled at some of the leaves exposing their skeletons; skeletons depicted on Minoan coins from 5,000 years ago. I can hear sheep and goat bells from down on the plain and the air is clear, cold, and mineral: it feels what it is—early autumn, with everything preparing for the rain and deep snow of winter.

Behind me on a rock is an Adonis Garden, an area of shallow soil with little hawkweeds and a tiny miniature onion, *Allium callimischon*, flowering in a drift of pink fragility. These patches of green are obvious downslope of every tree, like a slipstream or a wake—imagine the scree as a river and each tree as a post with silt building up in its wake. Some of the trees, the spiny kermes oak or prinos, the ancient gnarled cypresses, even the little nibbled junipers, are evergreen, but ambelitsiá is related to elm trees and is deciduous. The time it takes for it to come into and out of leaf is a period in which there is a large

---

* Acidic soils tend to contain high amounts of iron and aluminum ions, which also bond strongly with phosphate, preventing its uptake.

recycling of carbon: the leaves fall to the ground and break down to make organic acids, which contribute to the weathering of the limestone they grow on and help the tree to extract phosphate. The roots, particularly those shredding their tips as they move through the soil, also contribute organic acids. This is the reason that the Omalos plain I was looking down on is round—once upon a time it would have been a triangle, where three faults had pulled apart in different directions and the central area had sunk into the void, but organic acids coming down from trees on the sides of the faults had helped to dissolve the limestone and smooth the outline of the plain.

Zelkova abelicea has a warty, shaggy appearance, partly because of the gray bark, which flakes off in great satisfying vertical slivers, like slices of fridge-cold butter, to reveal the paler yellow-gray bark underneath. The bark is laden with moss and lichen—presumably why it flakes off so often, so the weight of the snow in winter doesn't bring the tree down. The lichen, too, contributes acids to the weathering effect on the limestone. The particular tree I'm sitting under looks old—I suspect it's a hundred—although it could be five or ten times older. At some point it lost a branch, possibly as a result of the weight of snow, and so the smooth area of the trunk, about as tall as I am, has a thickened patch at the top where a hole shows the loss, and the bark has thickened around the split that reveals gray twisted heartwood. Possibly, it was pollarded to make a shepherd's crook—walking over the peak of Mount Melidaou from the head of the Samaria Gorge to the head of the Eligias Gorge I met three shepherds, all of whom had ambelitsiá staffs in the back of their Toyota pickups.

The tree is rare because it only lives on Crete, and then only at high altitudes—above 4,000 feet. Across most of Crete the soil is too poor and thin for ambelitsiá to grow larger than a shrub. The biggest trees, and those most likely to reproduce, are down on the circular plain itself, taking advantage of all the soil that has been weathered out of the mountain over the years by the other members of its species. Having made its hard-won branches, it sprouts little leaves and twiglets along them, giving it a knobbly and slightly scruffy appearance. The nobbles on the emerging bark are therefore lots of shoot apical meristems, like little boles from which new directional branches can develop. Under the ground the roots do the same, branching fast from countless new buds, hunting phosphate through cracks and crevices, and enabling the ambelitsiá on this mountainous oasis on an island to find everything it needs to become a tree.

☞

At the other end of the scale, trees have had to develop ways to shape acidic soils in order to extract phosphorus from the organic molecules that have bound it up. At low pH (i.e., in acidic soils), soils have greater amounts of aluminum and iron, which form very strong bonds with phosphate; trees on old soil often acidify the environment.

The rare duraka, or "the stiff one," as it's called by the indigenous Baré people, grows only in one small area in Brazil, along the Curicuriari River, a dark, acidic tributary of the Rio Negro near São Gabriel da Cachoeira, under the dramatic jagged peak of a red granite outcrop, the Serra do Curicuriari. The tree grows a magnificent trunk over 160 feet tall and 13 feet round, sometimes buttressed, with deeply grooved flaking red bark, heavily colonized by thin lichens. In 1935, Adolpho Ducke, the unparalleled classifier of the Amazon flora, put the tree in a genus all its own, *Aguiaria*, naming it *excelsa*, meaning noble or lofty, indicating a tree that emerges beyond the canopy. The dry seed puzzled him: "I am not aware of a similar seed chamber in any other plant species," he wrote.[6] Unlike so many of the magnificent fleshy fruits of the Amazon it is dry and brown, with five valves on the exocarp that open like wings so that it spins gently as it falls from the tree canopy. When it lands, the five valves stand like the feet of a futuristic spaceship or a bacterial phage, holding the seed up, and under their protection the seed germinates—still well above the ground—and sends a root down for water. This allows the seed to land lightly on the very deep leaf litter, a blanket made by the tree for its progeny, and root.

The tree's rarity, and the fact that it is confined to such a small patch of the Amazon, is probably related to its seed shape: the five-winged helicopter doesn't allow the heavy seeds to travel as far as a juicy camu camu fruit swallowed by a large fish or a red-and-black huayruro seed spread by a bird. I sheltered under the trees here in a storm of magnificent proportions, and although I was soaked through, barely a breeze made it to the forest floor. Instead, the seeds are a mechanism for duraka seedlings to obtain enough phosphorus. The young plants, straight seedlings with bright-green pairs of leaves with opposing drip tips, are everywhere across the forest floor, on dead logs and suspended in lianas. Duraka trees are surrounded by legumes, trees that fix nitrogen with the help of bacteria, but phosphorus, which junipers can dissolve out of the limestone, is desperately rare. Brushing aside the leaf litter you get down to white sand, almost pure silica, mixed with a little plant matter that rots almost under your eyes. The ability of the seeds to land on the dead leaves of their parent tree and not be swamped in the almost knee-deep leaf litter gives them an advantage in obtaining enough phosphorus.[7]

Rivers can tell us much about what is happening along their banks. The Rio Negro has an oily black sheen, the Rio Solimões is cappuccino-milky, and their differences in pH and temperature are stark. Organic acids released by trees like the duraka of the Rio Negro lower the water's pH to 4, and because the weathered granite and sandstone of its catchment in the Guiana Highlands* in Colombia and Venezuela are some of the most ancient rocks known, made from the sand dunes of 1.7 billion years ago—long before life on earth—trees filter out all but a vanishingly small amount of the available nutrients, leaving only the humic acid of their efforts, which gives its honey-peat warmth to the water. By contrast, the young rocks of the Andes provide excess sediment to the Rio Solimões— more than enough to neutralize the organic acids released by the palms and cedro lining the Rio Marañon in Peru or the thick forests further down the Rio Solimões and provide rich nutrients to trees all the way down the river.

Dust from the rest of the world, particularly the Sahara, blows in extra phosphate, boosting the Amazon's productivity, but across all

---

* Part of the Guiana Shield, a pre-Cambrian formation that supports some of the areas of highest biodiversity in the world.

2.6 million square miles (nineteen times the size of Germany) phosphorus is the limiting factor for growth.★

When phosphate deficiency strikes in the rainforest, there is little plants can do apart from diversify to create worlds within worlds. Roots and mycelium spread through the shallow earth at the surface, fighting for phosphate at high concentrations from fallen leaves and other detritus that exude phosphatases. The root tip senses concentration changes of exudated primary metabolites and translates these into signals to modify root growth, allowing the plant to balance its development above ground with the resources available below. In tropical rainforest the intensive turnover of organic matter means that fast access to phosphate in leaves is necessary. Across old-growth Amazon forests, fine roots dominate in the first 12-inch depth of soil, and nutrient levels tend to be very low. Nutrients can easily leach out of topsoil, and so deep roots are sent down occasionally to bring ions up again, particularly in the case of potassium.

I had no way of seeing the roots of duraka, but on our way from the river to look at them we passed a towering root plate, taller than me, of an old, big Angelim Vermelho tree that had toppled, scooping out a wide semicircular sand pit of perfect white sand. The tree had fallen perhaps two months ago, said my companions, but already the trunk and roots were swarming with green, and young trees were shooting up to the light. The sand where the tree had grown was completely bare and white as a bone, apparently totally denuded of nutrients and almost pure silica. The traces of one or two deeper roots, sent to scavenge below, were ghostly imprints. Constant trade-offs are always taking place: investing in leaves for the moment, rather than in, say, woody tissues and fine roots. Fine, disposable roots will race to grow fast and straight in the direction of detected phosphate or calcium—a dead animal, a fallen tree. They will normally die back quickly once the desired mineral has been swiftly absorbed,[8] but live longest in tricky areas—acidic, sandy, or water-stressed soil—and

---

★ In the past there was a suggestion that megafauna had an essential role in spreading phosphorus by eating minerals in bulk.

will race over the top of the soil, sending occasional deep roots to scavenge for rarer minerals such as potassium.

Perhaps because of the piled-up pressure of competition, Amazon trees are particularly resourceful in the ways they use minerals that most trees leave in the ground or find toxic. Alfred Russel Wallace, whose codiscovery of the mechanism of natural selection with Charles Darwin was fueled by his extensive collections in the Amazon basin, tells the story of the pupúnha palm, whose elegant trunk is covered with painful spines (I speak from personal experience).[9] Wallace had captured some unfortunate parrots on the Rio Uapés, a tributary of the Rio Negro. Their "objections to any restraint upon their liberty caused me much trouble," he writes, and they bit through the wooden bars of their cages in a matter of hours. He was beginning to despair when a local man decided to help. "One of my Indians recommended me to try Pupúnha, assuring me that if their beaks were of iron they could not bite that." Wallace followed his advice and found that the poor parrots' "most persevering efforts now made little impression." In common with many Amazon trees, the palm's trunk is reinforced with silicates, packed into the cells of the xylem parenchyma. Silicates make up 95 percent of the granite at the source of the Rio Negro, and as well as being used for defense seem to be used by some plants as protection against the acidity of the river and the poisonous and common aluminum ions that can damage some plants.

A little further down the Amazon, the beautiful Vochysiaceae family of trees* takes up aluminum to use as a weapon, concentrating it in the most edible parts of its crisp, green leaves and purple or yellow flowers. Their roots will mine earth almost 30 feet away in search of aluminum and will hyperaccumulate the metal regardless of its availability in the soil. Such cycling of metals up to the surface fundamentally alters the vegetation around them. Ironically, they grow across the areas of some of the best bauxite deposits in the world and are threatened by human miners of aluminum.

In the absence of the conventional minerals needed to build the structures of life, other trees mine unlikely metals to use as replace-

---

* Particularly *Qualea grandiflora* and *Qualea parviflora*.

ment. Brazil nut trees, towering giants emerging from the forest canopy, mine selenium from the earth and transport it 200 feet up their trunk to replace scarce sulfur in the amino acids methionine and cysteine in their developing nuts. Replacement is hardly the word, as selenomethionine and selenocysteine work perfectly and poison unhelpful predators, such as humans, when they eat too many.* In short, the extent to which trees use every element they encounter deep underground would probably amaze us . . .

These changes are hard to quantify and can seem beside the point. Is it *shaping* soil to extract minerals from it and concentrate or weaponize them? Do these rare trees *really* have an impact on a global scale? For a human world convinced it is responsible for the Anthropocene, these molecular changes seem tiny: a satellite image will show the vast red scars of a bauxite mine, but not the tiny variation of green caused by a flourishing member of the Vochysiaceae family mining aluminum and weaponizing it in its leaves and flowers. In other areas, such as New Caledonia, where the nickel-miner *Araucaria laubenfelsii* is threatened by humans following suit, it is once again only the scars of human impact we can see—the tree is just green. In the snapshot of human time, local-level cycling of rare minerals seems a parochial shift in the world's minerals, an infinitesimal change in a grain of sand. If we take into account the fact that trees cover 30 percent of the world's landmass, however, and that they are the major route through which minerals get into the food chain, then even a quirky relationship becomes important. The rest of terrestrial life might end up depending on it. Particularly in times of chaos.

We have already mentioned the great shift of the Cretaceous-Paleogene boundary, 66 million years ago, when pollen from all over the world was replaced in the fossil record by fungal spores. It is a time characterized by the death of most of the world's forests, and the trees that survived struggled against darkness and changing weather patterns. When the world came out of the other side, angiosperms began to spread, and it was only after the Chicxulub asteroid

---

* Eating more than three Brazil nuts a day is not recommended as it might expose you to dangerously high levels of selenium.

that they started to dominate, outcompeting the gymnosperms until they occupied 90 percent of the globe.

There are many competing theories as to why this shift occurred. For years it was thought to be simply a matter of sex—the flowers of the angiosperms allowed trees to reproduce more successfully than the pollen dumping of the wind-pollinated gymnosperms—but other factors are now thought to be more important. One seems to be that, while gymnosperms worked to get what they didn't have, angiosperms started to work more with what they did. In other words, while gymnosperms were eaters of rock, angiosperms were competitive recyclers—more efficient but less fundamentally original. Hard as it is to prove anything in the muddy world of the underground, it seemed that angiosperms were the AI critic to the Proust of the gymnosperms, using their many and finer and better-directed roots to grab and resynthesize what had already been brought up and synthesized over thousands of years by the thicker deeper roots of the gymnosperms. "I follow the tortuous path of roots bursting the earth," wrote Clarice Lispector, the Ukrainian/Brazilian writer and pioneer in experimental novel-writing. Angiosperms followed the tortuous path of gymnosperms.

Recycling leads to intense competition underground, with trees weaponizing soil against each other. The Shennongjia region of China, part of the Dabashan Mountains, is a convoluted massif covered in the most biodiverse deciduous forest in the world. The fossil record shows a deeply stable area in which tree populations change and develop slowly over millions of years. The region is named after the mythical farmer-emperor Shennong, who in the fourteenth century BC is said to have invented tea drinking, the plow, and the hoe. A story in the Chinese *Epic of Darkness* describes him using a rattan ladder, a *jia*, to climb to the mountaintops, which, when he dropped it, clattered down and transformed into forest.

In 2020, an experiment performed by Lijuan Sun at the Institute of Ecology in Beijing showed that angiosperms were competing fiercely via their fine root exudates. His team tracked eighteen interlinking tree and shrub species in the area and showed that they were in a biochemical dance with occasional rapier thrusts. Rough coalitions could be drawn between the evergreens and the deciduous trees, between more closely related trees, and trees at different heights, but beyond that they turned on one another. On the evergreen side were the laurels, the persimmon, and the rhododendrons, with the semi-evergreen viburnum mediating. On the deciduous side, the maples, *Acer davidii* and *Acer palmatum* as well as *Acer mono*, coexisted with the oaks, the beautiful *Cyclobalanopsis oxyodon,* so-called because its acorns peep out of a round, fused cup, built up like magma from a lazy volcano, out of which the acorn peeps like a traditional Chinese cap, and *Quercus serrata*, with large leathery leaves and deeply furrowed bark. These slogged it out with a large-flowered magnolia and a sorbus. The eighteen-way competition is more than mudslinging: each tree is hoping to control, through the release of acids and enzymes, which metals and inorganic ions will be available in the soil.

In other parts of the Dabashan, gymnosperms remain aloof from competition in stands of one or two species. Among the bizarrely weathered granite stones of the massif, stands of *Cryptomeria* (literally "hidden parts," because the flower parts are hidden in the foliage) paint stripes of shadow-green along misty hanging valleys. Although some stands are 1,000 years old, they're believed to be a planted introduction from Japan, where trees over 2,000 years old, like the Jōmon Sugi, grow to be huge. Most commonly, however, *Cryptomeria* is planted in great monoculture plantations for timber and wood. Intrigued by the success of the tree for forestry in Japan, China, and the Azores, where it was introduced in the nineteenth century, scientists researched the soil beneath *Cryptomeria* and found that soluble calcium and magnesium were much higher than in other plantations, allowing the trees to grow faster and straighter in a shorter amount of time. The trees were exuding so much acid in warm, wet conditions that they digested the rock under them more efficiently than anything else and killed all competition.★

★Jōmon Sugi is described by the wittiest of tree men Thomas Pakenham as "a grim titan of a tree, rising from the spongy ground more like rock than timber, his vast muscular arms extended above the tangle of young cedars and camphor trees" (Pakenham, 2003).

Cryptomeria have been found in fossils in Derbyshire and Iceland and Italy. As with the ginkgo and the dawn redwood, their fore-bears were once common across the northern hemisphere, which means that these trees were shaping the majority of the earth with their intense acid flow for millions of years. Gymnosperms have been called the archetypal stress tolerators, and when angiosperms started to scavenge off their deep-rooted soil and outbreed them, they were pushed into strange places. One place where they still flourish is New Caledonia, an archipelago between Fiji and Vanuatu that hosts forty-three endemic gymnosperms. New Caledonia is a fragment of Gondwana, the ancient supercontinent that started to fragment about 180 million years ago, and much of its rock is ultramafic: volcanic rock with a very high magnesium and iron content. It can also have extremely high levels of nickel, cobalt, chromium, and manganese and tends to be acidic. Only the best-adapted trees, such as the *Araucaria* species, gymnosperms with scales and strange ungainly branch-ing patterns whose most famous relation is the monkey-puzzle, can carve out a niche here, and they are starved of nitrogen, potassium, and phosphorus.

No one has yet solved the mystery of how *Araucaria* can maintain a foothold on these poisonous soils, but one of my lecturers at Oxford, Andrew Smith, was pursuing a line of research that offers a clue. He was working with a very small cress called *Alyssum* (now *Odontarrhena*) *lesbiacum*, so-called because it grows only on Lesbos, the Greek island famous for its ultramafic serpentine soils. I know these same soils from the tops of mountains in western Turkey, where they host clusters of tiny yellow bells and gray-green leaves, *Fritillaria ser-penticola*, a tiny bulb endemic to these soils and these mountains. Ser-pentine soil is thin on top of the rock and hard to climb up; it slides downhill, getting under your fingernails and into your shoes, and is a beautiful red-green-brown that can glimmer in the sun like snake-skin, and lump out of a limestone hillside like a snake's back.

Most plants that can tolerate toxic soils can discriminate against and not take up toxic metals from soil. The cress that Professor Smith studied, however, took in nickel from the soil and stored it in its shoots above ground to a concentration of 4 percent—debilitatingly high in most circumstances. It does this by producing free histidine, a positively charged amino acid with a nitrogen-containing ring, which can bind and detoxify the nickel ions as soon as they enter the root system. A tree from New Caledonia, *Pycnandra acuminata*, accumulates so much nickel in this way that when it is cut it bleeds sea-green latex from its stem containing 25 percent nickel in dry weight. It seems these hyperaccumulator plants do this not because it in any way eases the difficulty of living in toxic soil, but to use the metal as a defense against fungi and bacteria. By increasing the nickel concentration in nearby soil through leaf litter, they protect themselves against herbivores and even competing plant species as well.

Having ground down the soil, why do trees fill it with so much carbon? Nowhere is tree rock-mining more visible than in the Ziarat Juniper Forest in Pakistan.* From Quetta, an hour's drive into the arid mountains that form the border with Afghanistan takes you to 425 square miles of the world's oldest and toughest trees. These are *Juniperus excelsa*, taller and more impressive than the juniper of gin and tonic fame, scattered across the otherwise bare hills.† Legend has it that they walked across the mountains from Afghanistan and settled here on the bare rock, breaking it up to make a home for a wandering Sufi saint and his people. It's easy to see where the legend comes from—far up on the hillside, I saw one tree growing on a huge fallen chunk of stone, with a root almost as thick as its trunk heading straight down, cracking the rock perfectly in two.

These trees mine rock quite literally for nutrients to survive on, but they make soil by putting carbon back among the fragments of rock they have broken up. Their fine roots are constantly probing the rock around them and then dying back a little, sacrificing some

---

*This also happens to be where the founder of modern Pakistan, Muhammed Ali Jinnah, spent his final years, in an austere and beautiful wooden house with a green-painted veranda.

† The juniper used in gin is *Juniperus communis*, a shrubby, winding plant.

complex carbohydrate to encourage the emergent ecosystem around them and make a little soil. Then they take advantage of the mutually created environment. The older the trees, the more carbon they commit to the soil, and you can see this in the vast roots of the gnarled trees said to be over a thousand years old, flaky-barked at ground level, looping over rock and through earth, twisting with the torque of centuries. In one place, the mountain road had cut a slice through the roots of an old tree, and there they were, looping horizontally and readily through the soil and in a fracture pattern like the punch from a pneumatic drill inching more cautiously through the splintered rock. Next to it a tiny juniper seedling had put its own root—much longer than its trunk was high—into the soil of one of the fissures made by the parent tree.

In the 1920s, a Swedish botanist called Henrik Lundegårdh started to investigate the effect of plants on soil and found they were adding an extraordinary amount of carbon to it. Centuries of thinking about the soil had ignored the role of carbon, with all the great natural philosophers—Aristotle, al-Kindi, and Confucius—attributing its effects to different causes. In part, this is because trees put carbon into the earth in three different ways. The first is the most obvious: leaf and twig drop rots down into humus as a result of the actions of a million microorganisms. The second is the gentle rotting of roots in the ground, and the third, as we have seen, the active pumping of carbon-containing goo into the earth through the roots. All three processes make up soil organic carbon, long chains of carbon that, having been joined together by plants, are trapped in the earth, freed only through decomposition by microbes, or occasionally by the plants themselves.

Lundegårdh and the other botanists who started to piece together the carbon fluxes that occur through the soil were working at the level of individual plant metabolism and then extrapolating out to a global level, but because of the need to be rigorously experimental they worked on wheat and cress—small plants with small roots. Trees, both larger and older, trap exponentially more carbon into the earth as they mature, which means that soil organic carbon is a black pool of strength underpinning woods and forests, hedges and open-grown trees, essential in the fight against climate change. It also means, of course, that deforestation has the potential to release carbon at a cataclysmic level. The junipers of the Ziarat Juniper Forest are weaving carbon chains through the rocks, but were the trees to be lost, the carbon would degrade and slowly seep out into the atmosphere.

Would you risk your life for a tree? Would you risk your life to fight climate change? Would you risk your life to complete a spreadsheet calculating the carbon—particularly the soil organic carbon—locked away by 425 square miles of mountainous tree cover? When I said I would like to visit those junipers in Quetta, my host in Karachi sounded a little worried. It was, after all, only 125 miles from Kandahar in Afghanistan. But soon he emailed back with his habitual courtesy to say that we could go and would be hosted by the Balochistan Forestry Service.

At the airport Dr. Niaz Khan Kakar and his team picked us up with an enormous wreath of red and white roses, huge smiles, and a big armed guard, which I thought was excessive until we drove out of the city to the forestry headquarters. In one village it was market day, and the streets were blocked. The few women were veiled in burqas and all the men had AK-47s. Some were also carrying submachine guns and pistols. One man had a knife stuck in his belt; another was touting an RPG. "Nonsense!" I said robustly to my companion. "Just because people are carrying weapons doesn't make them the Taliban!"

"The Taliban?" said Dr. Niaz, who was rapidly becoming a friend. "Yes, these are the Taliban; last year they killed my brother, who was a judge. But you could walk through that market now and be fine. You would have to establish a routine here before they would harm you, and even then, the third time you walked through, or the fourth or the eighth, they would probably just kidnap you for a ransom—*poof!*—over the border to Afghanistan. You look so American and so rich . . ." He was laughing at the last sentence, and I thought ruefully about my bank account and my long coat, faded and shredding, borrowed from my mother for decency's sake.

But later that evening, as we sat in the twilight under huge chinar—oriental plane trees—wrapped in blankets listening to musicians in exile from Kandahar singing of cherry trees and blossom, of girls who go to the well and then dally under the willow trees in the moonlight, I realized something of what it meant to be part of this organization who risk their lives every day for the Ziarat Juniper Forest. Many are women. Maria was from Quetta's embattled Catholic community and worked defiantly bare-headed. Gul was Hazara, also a minority persecuted by the Taliban in Afghanistan, and brought one of her eleven siblings with her each day. Babrak had been a member of the Taliban before joining the civil service, but had learned English from Jane Austen and, like me, knew vast tracts of *Persuasion* by heart. He recited it as he drove us along the mountain roads with supreme skill at terrifying speed. Blowing a cloud of hashish out of the window, he said, "None of us want to be in calm waters all our lives." Clinging to the door handle, I was only partially inclined to agree.

Valiant tree lovers from all areas of Pakistan, the Balochistan Forestry Service face down threats galore on behalf of the trees—not just the Afghan Taliban on one side of the border and the Pakistani Taliban on the other. They also tackle more complicated moral questions. As we sat under the moon and talked about this, Dr. Niaz Khan Kakar explained the complexities. The Taliban smuggle drugs and other goods across the mountains here. With opium no longer abundant, the major drug is methamphetamine, which is easily synthesized from ephedra, a small spiky plant that grows across the mountains.* They will think nothing of vandalizing a tree so that it dies, blowing it up to make a road or cutting it down and selling it for a bit of extra income, and there is a vast area of old trees along inaccessible and steep mountainsides they can target. Even so, trees can be simply—if dangerously—defended with guns. The harder problem comes from the simple flourishing of local people: the friends, neighbors, and family members of the forestry service who have themselves protected and used the trees for hundreds of years. Traditionally, juniper branches are used as fodder for animals,†

---

* A fascinating, very early branch off the tree of life, related to gymnosperms and little else.

† This fodder is most highly prized when mistletoe has grown above the branch, after which the foliage is much sweeter and more palatable for the livestock. This is because the mistletoe circumvents the protective mechanism of the tree—turning its bitter tannins to sweet sugars.

the wood for fires, and the brown flaking bark as roof shingles for houses. As the population grows, these traditional uses of the trees are becoming unsustainable.

The solution the forestry service has found is to sell carbon credits. The organic carbon that trees have put into the ground is calculated, and North American companies buy what are essentially carbon offsets. The money is used to protect the trees and provide alternatives—oil-fired cooking stoves, sustainable fodder, or simply additional income—for the local population. It works because the forestry team on the ground is endlessly out and about in the mountains, chatting, chatting, chatting; acting as a social support network as well as drone-flying, LiDAR-operating scientists.

The forestry service has meticulously researched the exact amount of carbon trapped in and around the trees, and some of the results have been surprising. The carbon in a tree above the ground is simple to compute: in twigs, leaves, branches, and trunk you're looking at a wondrously shaped and colored lump of carbon, so what you see is what you get. In the Ziarat Juniper Forest you get shapes of extraordinary complexity: tall, cone-shaped spikes reaching to the sky; on the lip of a cliff a tree like a crab clinging with a multitude of crooked hands to its precarious crag, leaves hanging on only on the underside of the branches. A little further down the mountain the miraculous "Allah" tree has a straight trunk for the first *alif,* and then one immense trunk looping over horizontally to provide the ligature for the *lam, lam, ha* branches growing up vertically from it. Twenty years ago, to capture and measure the carbon dioxide (CO_2) in this assembly you would have had to cut down, weigh, and possibly even burn the tree. Now you can simply 3D-scan the tree above ground using a drone or a phone, and you have the tree's mass, approximately 18 percent of which will be carbon.* It's an example of carbon credits

---

* Below ground, calculating the carbon that a tree has put into the soil and trapped there becomes more complicated. As well as roots (made, like branches) of carbon molecules chained together into organic matter, an invisible framework of carbon chains—the soil organic carbon—will have been released into the soil by roots as part of root exudates, and this invisible framework must be calculated. Trees in the Ziarat Juniper Forest grow far apart, but the amount of carbon in the soil here is

doing what they were designed for: polluters paying to keep carbon in the ground in places where it might otherwise be lost. But it still encapsulates a fundamental tension between humans and trees: even at our most sympathetic, we find it almost impossible to live in sync with the productivity of a tree.

We had already had an exhausting couple of days, and I was starting to nod off. When I jerked my eyes open a ring had formed round the fire. Our guards, who had been behind us, murmuring in the darkness as the music played, had put down their guns and linked up in a circle with our hosts. To the screel of the musicians, laughing as they played faster and faster, the guards were dancing, leaping higher and higher as the drumming got fiercer and fiercer. That night I dreamed that the junipers were growing to the sound of the drum, their branches like angular arms, their roots like knees, their feet striking the earth with persistent patience.

Rhizodeposition, the carbon added to the soil by roots, has many purposes. The pressure that a fine root must assert against rock or even soil can reach 100 pounds per square inch. That's approximately four times less than the pressure of a 4-inch stiletto heel of a Louboutin shoe, but five times the pressure of a horse standing on your toe. Like a repeatedly blunted tunnel borer driving a road through a mountain, the pressure destroys the cells of the root tip, so a special root cap of cells is created. As the root moves forward, these cells slough off, and to minimize this wasteful process and protect the rest of the root from damage the tree produces mucilage, a layer of long and short chains of sugar that can glide over one another even under high pressure and act as a lubricant. The makeup of this sugar-slime lubricant secreted by the roots is meticulously

---

surprisingly high, given that the density of roots is low. Forest scientists hypothesize that long-lived pioneer trees, like these junipers, have purposefully evolved to put more carbon into the ground at first, thus establishing a niche that will support them in its turn.

adapted to the crystal size of the rock or the grain size of the soil. It's insoluble, so it can't be washed away by a fierce mountain rainstorm, and sticky, preventing pathogens from getting through. Deposited in the soil it encourages weathered rock particles to aggregate, trapping water and preventing desiccation, and the cells the root leaves behind stay alive for a time even after they have left the root, continuing to secrete carbon and taking up poisonous metals as a sacrificial final act.

The roots are also exuding organic acids and enzymes to extract nutrients from the soil in soluble form and feeding bacteria and fungi, often in exchange for nitrates or phosphates. All around the world junipers employ similar tactics to shape hard earth and tricky places. The junipers at Gish Abay in Ethiopia, one of the sources of the Nile, directly exude phosphatases and organic acids, which help them to liberate the nutrients they need to grow. The surrealist skeletons of sea junipers (*Juniperus macrocarpa*) in Crete and its satellite island Gavdos grow on shifting salt sands, under intense beating light reflecting off the Mediterranean. One tree, which sticks right out into the Mediterranean, is the southernmost big tree in Europe, and it is only root exudate that has allowed it to anchor itself so firmly in the shifting soil. Elsewhere in the world you find the coppiced *Juniperus phoenicea* growing out horizontally over ravines, the large-fruited *Juniperus oxycedrus* subsp. *macrocarpa* finding a foothold in moving sand dunes, and a profusion of Mexican junipers whose taproots can extend over 23 feet vertically down in search of water. In the Guatemalan highlands, almost exactly on the opposite side of the earth and nearly 25,000 miles from Pakistan, *Juniperus standleyi* grows on equally bleak rock faces in the mountains and is also used for roof shingles and firewood. Whether on sand or rock, at sea level or very high altitudes, these trees are masters of turning rock into dust and dust into earth.

Junipers might be particularly good soil engineers because they evolved during periods of sweeping drought, and this is now leading to the development of drought-resistant varieties. Some of their closest relations are found in relict populations in warm, wet areas around the globe. The most famous example is the alerce, which can grow to 230 feet and live to over 3,000 years old. Darwin recorded one with a trunk circumference of 41 feet. The alerce grows in humid forest in Chile that reaches from the mountains down to the sea, but fossils of a close forebear have been found in Tasmania, suggesting it once grew across the southern hemisphere. The vast ahuehuete or Montezuma cypress clings on in Mexico and Guatemala, and the largest example, a 630-ton monster (three times heavier than a blue whale), is watered by the town of Santa Maria del Tule, where it takes pride of place in the town square. And on the other side of the world, the critically endangered Vietnam swamp cypress lives in swamps and ponds in subtropical China and Vietnam, where it is used to baffle the sediment of rivers and prevent soil erosion.

The swamp seems to have been the original habitat of the common ancestor of cypresses and junipers, but in 2008 a paleobotanist called Ignacio Escapa was digging in Patagonia when he found imprints of cones in seed and prepollination, twigs, and branches, which turned out to be those of a member of the Cupressaceae that had adapted to dry conditions around 200 million years ago.[10] Like the cypresses and junipers around today, the fossil had cones and flattened, ever-green scales rather than leaves. The fossil implied that the Cupressaceae species, having evolved early and been successful around the world, survived the warming and drying of the late Jurassic in various forms, but then faced a choice: adapt and spread or stay in their niches. Those species that are now spread across the dry areas of the northern hemisphere were the dynamic ones, adapting fast to put a girdle round the earth.

And one adaptation was counterintuitive: under drier conditions they exuded more carbon, not less, from their root systems. Indeed, thirsty trees universally exude more carbon. The supposition of the Pakistani forestry commission, that pioneer trees living in hard, dry conditions work hard to make the soil richer, is backed up by research conducted by scientists in Israel, who have measured the levels of fluid that cypress (*Cupressus sempervirens*) and mastic (*Pistacia lentiscus*) trees produce through their roots under drought conditions. When the trees are stressed by lack of water or by heat, they use some of the precious liquid they have left to increase the amount of carbon they put into the soil via their exudates. In fact, cypresses will put eleven times more fluid into the ground via their roots in the hot summer months, while the mastic trees they grow alongside will increase the liquid they exude four times. The liquid released typically includes sugars, amino acids, and organic acids— the first products of plant metabolism—and the more complicated carbon chains that trees create after that.★ What the Israeli research- ers showed was that trees in this forest reinvested in the soil, feeding in extra carbon in drought-stressed months to bolster the bacteria that provide the majority of the phosphatases and organic acids trees need to obtain essential nutrients, particularly phosphorus and nitrogen, from the soil.

It used to be a mystery why nothing would grow in the bare earth of the Klamath Mountains in California. Finally, it was solved. The soil was poisonous, and the poison was acid, leading to a soil pH of 4.5, the result of nitrates made by bacteria millions of years ago in the deep ocean seeping out of the mica-schist rock and dissolving in water to make nitric acid. By contrast, in the pine forests across most of the mountains nearby, the nitrogen in the rocks was not enough. Like animals, trees need nitrogen for every one of the enzymes that acts as a shape-shifting biological catalyst, particularly Rubisco, the huge, lumpy enzyme that fixes CO_2 from the atmosphere. Without nitrogen in the soil, trees stop growing.

Inorganic, soluble nitrogen in rock can provide about a quarter of the nitrogen that trees need. Lightning is another spectacular way of making nitrogen bioavailable for trees, slamming nitrogen gas from the atmosphere into hydrogen or oxygen. Most of the nitrogen found in the soil before 1910, however, was made by a symbiotic part- nership between bacteria and trees, in which trees imitated the deep

---

★ Increased soil organic carbon increases cation exchange capacity (the number of positively charged ions that can be retained on soil particle surfaces), allowing trees access to more micronutrients and nitrogen.

ocean, the anoxic home of the diazotrophic bacteria.* These bacteria
had turned nitrogen into ammonia for 3.4 billion years, but rather
slowly and never in the presence of oxygen. By the time trees were
dominant in the Carboniferous period there was a huge nitrogen
shortage, but the problem was fixed by a group of trees called cycads,
feather-leaved plants sitting on a lumpy, branched set of roots that
can be shredded and reformed at will. These roots are called coral-
loid roots because they have nodules and a pink tinge like coral; they
produce special oxygen-free areas designed to nurse nitrogen-fixing
bacteria.[11]

The cycads were the first nitrogen fixers, hosting bacteria that
fed nitrogen to the carboniferous. They flourished in the time of
the dinosaurs, but nearly all cycad species are now in decline, superseded by angiosperms—mainly the legumes, shrubs, and trees with
pea-shaped flowers like those on the white sand of the Serra do Curicuriari. It is energetically very costly to fix nitrogen; the fossil record
is full of plants that tried to become nitrogen fixers and eventually
became extinct through going energetically bust.[12] Those that hang
on are hugely influential enablers, allowing other trees to move into
inhospitable areas.

Tree planters know that alder is one of the keys to successfully
establishing trees on degraded waterlogged soils. Where the ground
is damp, alders nitrify and aerate the soil before anything else will
grow, and, just like the cycads did, they do it using bacteria, in their
case the bacterium *Frankia alni*.† The bacterium is such a powerful ally that from the first moment trees sense it near them they
start to manufacture mechanisms to protect it from oxygen and se-

---

* Since 1910 the Haber process has done this to produce industrial fertilizers for
agriculture.
† Whereas gymnosperms take up ammonium preferentially, angiosperms tend to
be poisoned by ammonium and prefer nitrate, which they can endure because a
special new gene (caused by gene duplication) allowed angiosperms to take up nitrate into their chloroplasts, and this lessened the poisonous effects of the nitrate
in the leaf. As a result of this mechanism, ammonium uptake is mediated by sun;
when there is a lot of sun, the more nitrate a tree will take up. Less sun forces trees
to hunt for ammonium.

duce it in. When I was spear-fishing in the alder-fringed Kalamos River in northern Greece, I was struck by the remarkable roots (in which alder houses its bacterial armies) fringing the river banks. More developed than those of the cycads, the roots look like stubby worms, with carmine bodies★ and white tips, and oxygen is kept away by the tree triple-wrapping the bacteria, first with tree-glue pectins, then by the formation of specialized cell layers covered in cellulose, and finally by a change in the metabolism of the specialized cells that house them. Moreover, the bacteria-housing roots grow underwater to avoid oxygen, and are red because they are full of hemoglobin,[13] the same iron-carrying protein that binds to oxygen in our blood. The tree was doing all the work, not just in the case of hemoglobin, but in every sense: supplying everything needed—protective structures, sugars and organic chemicals, water, proteins—all so the bacteria could fix nitrogen and supply it to the tree in a soluble form. Consequently, the alder trees can outcompete any tree on the river and supply nitrogen for the entire riparian ecosystem.

There is so much we still don't know about how roots shape the earth. Incredibly, no one knows what it is that dictates when roots change direction to head toward water. It seems to be a movement that is independent of any auxins, but it may follow vibrations. We still don't know exactly how trees found enough nitrogen during the Carboniferous, though fungi are likely to be part of the answer. And we don't know why trees will sometimes cooperate and sometimes compete underground. At the beginning of this chapter we talked of trees as blind explorers. The idea of acute, fast senses and responses is beside the point when it comes to trees, but the way they choose their direction is directly pertinent to how they shape the world. In the process they irrevocably shape human thought: in so many human

---

★ Walt Whitman talks about "the pink-tinged roots" of *calamus* in his "Calamus" poems. He appears, alas, to be referring to a different root altogether.

cultures the dark rhizosphere, or the even darker place beneath it, is the domain of death and hell.

For her doctorate my sister studied the tombs of Lycia in southern Turkey—limestone caverns honeycombing the cliffs with stiff, geometric frontages smashed open by tomb robbers. While she was tracing faded curse inscriptions I sheltered from the sun, marveling at the determination of the odd root working its way through the solid limestone of the tomb ceiling and gradually dismantling the tomb into dust. I remember one Byzantine cistern with a recycled block of granite holding up the roof; a fringe of roots from an oak above had erupted through the ceiling like a spray of mistletoe. The mystery of what these roots were groping for in thin air haunted me then and still does. Eventually I went and signed up for an archaeological field school in Belize, to spend time thinking about roots and the underworld.

Mayan culture believed in a world in which the magnificent *Ceiba pentandra* tree held heaven away from earth and earth away from the underworld, but also was a conduit through the world. *Ceiba* is a smooth-barked tree like a massive pillar—its roots were everywhere across my sister's dig. At Baking Pot, one of the least excavated sites, an entire palace wall had been consumed by the buttresses and roots of an enormous specimen that was patiently prising apart the hewn blocks of limestone. Elsewhere, broken Mayan pottery was strewn across the ground, the work not of grave robbers but of churning *Ceiba* roots. We followed a stream of cold air into the mouth of a cave that ran deep under the Cayo Mountains. In the shallower caves roots snaked down through the roof, ran along the camber of the cave, and then grubbed about restlessly among the thick fruit bat guano of the floor. Enormous chambers full of bats opened out next to smaller passages full of black charcoal and Mayan pottery smashed and burned, the contents sent to the gods. Jaime Awe, the head of the dig, who had been investigating the cave ever since wandering in as a teenager twenty-five years earlier, explained what the archaeological evidence had shown. Mayan civilization had entered a period of drought and famine, and as this disrupted the Mayan societies around the cave, sacrifices and rituals had become larger and more intense, while sacrifices of humans, particularly children, increased. Most of what we could see dated from this period of societal breakdown, when the community was desperately trying to restore fertility to its world by coaxing it out of the underworld. Archaeologists think now that population growth had caused deforestation and soil erosion of the region's thin soil.

For me the cave was quite near enough to hell. It was a dark maze, constantly branching, the air thick with smells from various caverns: dust from drier, crumbling rock, undecomposed bat guano, sparkling stalactites, thick layers of mud deposited after recent torrents, and occasional tree seedlings germinating from fruit dropped by the bats, growing up straight in expectation before hooping over and withering in the darkness. The stone, impossibly sterile, was carved by air and water, formations were abrupt and unsoftened, and every so often a spine-chilling slither of a rockfall would echo through the tunnels. After walking for about 1.2 miles in pitch black we saw a glimmer of light on the ground. Up a hill of mud and sediment was a sinkhole, a plug of pure green light where trees clothed in lianas grew up toward the light from 40 feet down, and a rare cycad had established itself in a stable nook.

I thought of the sterile, dark land of water and minerals and gray dust we were leaving. Without trees like the junipers driving their Louboutin-pressured roots through the soil, most of the earth would be a moonscape of bare rock. If a rock did crumble into dust, there would be no sticky carbon chains, rhizo-deposited by a root groping toward the center of the earth, to hold it together. It would just blow or wash away, as it did for millions of years before trees started to climb up and reach down. Phosphate would be locked away in an inaccessible form, deep in rocks that had never been worn away by organic acids or phosphatases. Nitrate, an essential component of our bodies as well as all life on earth, would be scarce and rarely available—a lightning strike here or there might fix some, but few bacteria could cling on in the absence of their supportive hosts feeding them carbon. As surely as a sinkhole, trees have illuminated the darkness of the earth and brought out of it green-drenched soil.

CHAPTER 3

Trees shaping fire

Why do trees evolve to burn?

*Just as all-consuming fire burns through huge forests on a mountain top,
and men far off can see its light, so, as soldiers marched out, their glittering
bronze blazed through the sky to heaven, an amazing sight.*

Homer, *The Iliad*

Abies equi-trojani, the Trojan horse fir,* grows above the plain of Troy
in Turkey. The mountain is also known as Kazdağı, literally Goose
Mountain, after the shamanistic Tahtacı Turkmen clans who live
there and venerate geese. The clans also venerate trees and were originally settled on the mountain in the fifteenth century by Mehmet the
Conqueror to cut down the firs for the timber needed to expand the
Ottoman Fleet. The trees grow slowly, making hard, strong wood,
but for the six centuries they were managed by the Tahtacı their
numbers were stable.

In the past thirty years, though, despite valiant conservation efforts, *Abies equi-trojani* has been on the decline. It seems a textbook
symptom of global warming: driven off the top of the mountain as
its former habitats get too hot[†]—until you map the movements of
the trees across the region. Scientists from the university nearby suggest that firs are becoming rarer because they are being burned off

---

*So-called because Coode and Cullen, two imaginative Edinburgh botanists
brought up on Homer, decided that this must be the tree that was cut down to
make the Trojan horse.
[†]Precisely what has been happening to *Abies nebrodensis*, a very rare fir, of which
only about twenty-one specimens remain in the high mountains of Sicily.

the mountain by the trees they live with, black pines, *Pinus nigra*. Recently the researchers found that there have been many small forest fires on Kazdağı, some caused by humans and others by lightning, after which the pines have been outcompeting the firs, regreening or growing back so quickly that they completely shade out any fir saplings, and driving them to extinction.

This seems an unfortunate coincidence, but the contrasting biochemistries of pines and firs make it clear that it is not. When pines experience water shortage the first thing they do is to close their stomata—effectively stop breathing—in order to retain water in their cells. As a result, biochemical pathways that would normally be producing sugars spin off to produce naturally flammable substances, terpenoids, hiking the chances that the pine will catch fire. If the drought continues and the lack of carbon dioxide drives photosynthesis itself to a halt, the trees start to produce shikimic acid, the forerunner to even more volatile alcohols.* Alcohols catch fire quickly but burn cool, so this alcoholic haze will ignite at low temperatures, massively increasing the chance of fire, but reducing the chances of pine seeds being annihilated by heat. Walking across Kazdaği I've seen pine trees that have been struck and killed by lightning but remain standing, hollow, taller than all the trees around them, retaining their resins and acting like a match in the next storm. Fir trees do the opposite. Their response to drought is to produce what is essentially a fire retardant, a thick gooey oleoresin that is very hard to ignite. The ecologist Oliver Rackham put it best: "Trees and other plants," he said, "are not flammable by misfortune but by adaptation."[1]

So why do pines self-immolate? Why would plants adapt to burn? In the pine tree example the evolutionary advantage is clear. Pine, when burned, can grow more than 3 feet a year, quickly springing up from seeds and shading out its competition, in this instance

---

* Broadly, small carbon chains burn cool and long chains burn hot. Alcohols, for example, often burn at a low enough temperature that you can hold them on your hand, whereas something like wood or coal burns hot because of its long carbon chains.

the fir trees. It's a way of turning a marathon—which the fir might win—into a sprint, which it won't. Add to this the structure of pine needles, found in clusters that fall to form a bouncy, aerated layer of kindling on the forest floor, and you see a tree turning fire to its own advantage. Under the pines, fire races across the ground, destroying parasites and the seeds of invading plants. The heat of the fire melts the resin that has sealed the pinecones shut, the pinecones pop open in the heat, and the seeds drop on the ash, which is rich in magnesium, potassium, and iron. This adaptation arises quickly among trees that live in fiery environments and is called pyriscence, literally "following fire." With any luck, the lightning that ignites the pine will also be the forerunner for a storm, heralding more lightning and new rains from water particles that gather around the wildfire smoke. When a lightning bolt strikes, it solidifies nitrogen gas from the atmosphere into nitrate minerals in the soil with each high-energy impact on the earth. Suddenly, from being in a bad position, a million new saplings are ideally placed to dominate the land they've fallen on, facing no competition and with abundant water and nitrogen.

In Paris in 1737, a competition was held to determine the nature of fire. At this historical juncture the actual nature of fire was still uncertain. Was it an element? A force? The theory of phlogiston—that an essence of fire was mixed in with, and then emanated out of, burnable substances—was still used to explain oxygenation reactions like burning and rusting, but although it had many advocates it was obviously full of flaws.* One of those who entered the competition was the grand and beautiful philosopher Emilie du Châtelet (1706–49), a prodigy and a savant who, by the time she died in childbirth at the age of forty-two, had made her name with works on philosophy, physics,

---

* Phlogiston theory was finally disapproved by Lavoisier in the 1770s, when he demonstrated that the mass of a gas was included in the substance, and the theory of catalysis, along with the redox reaction mechanisms published in the 1790s by Elizabeth Fulhame, disproved the phlogiston theory for rust.

and ethics. She produced a French edition of Isaac Newton's *Mathematical Principals* that is still used today, introducing an additional law of conservation of total energy by adding in equations for movement.* She was also Voltaire's lover and collaborator,† and she was ahead of her time in thinking about scientific method, defining the criteria of a sound hypothesis in the way we still would today. She entered the competition in 1737 secretly, telling only her husband, after disagreeing with Voltaire over the results of experiments they conducted together. Whereas Voltaire was convinced that fire consisted of particles that had mass and weight and obeyed Newton's ideas of physics, du Châtelet concluded from her observations that fire is not a material substance. She didn't win, but her essay was considered important enough that it was published in a special pamphlet by the Paris Academy, along with those of other competitors, making her the first woman to have a paper published by this bastion of science.

I couldn't find a translation of her work and, realizing I had to read fifty-three pages of mathematics in French, my heart sank, but *Dissertation sur la Nature et la Propagation du Feu* reads lyrically.[2] Emilie du Châtelet is a fluent writer. "Heat and light," she writes, "are the effects of fire that most strike our senses . . . but we must still understand that they do not always work together." She explained this through chilling experiments with glow-worms: "I immersed them in very cold water, and their light was not altered."[3] She came incredibly close to identifying something that Lavoisier would in due course prove in 1790: "Perhaps it may even be that the kind of rarefaction that operates with air and water is caused by fire itself." Fire, she points out, is the cause of the internal motion of the particles of all bodies we think of as fire as detected by our senses, that "depend on our existence and the manner in which we exist; for a blind man will define fire as that which heats, and a man deprived of feeling, as that which illuminates."

---

* It is this law that in Chapter 1 calls into doubt the biotic pump-mechanism theory of forests sucking water away from the sea.
† His heartbroken description of her to Frederic II of Prussia was that she was "*Un grand homme qui n'avait de défaut que d'être femme*"—"A great man whose only fault was that she was a woman."

We are dealing with blind men, trees, and we must try to experience fire as they do: as a force that cannot be fled from in an instant or seen in the smoke of a distant mountain, but can be sensed through the tantalizing whiffs of carbon monoxide, formaldehyde, and acrolein in the air taken into stomata or repeatedly experienced as sap boiling in xylem and phloem, sterile wounds etched deep into heartwood, and possibly as the death of everything above the ground. Sometimes trees will experience fire and its aftermath through their leaves, as an increase in light; through the shoots, as a reduction in predators; and through the roots, as a fresh draft of nutrients. As proteins are flexibly remodeled to express the genes to respond to this, so these epigenetic changes are passed on. As we will see, trees can change their entire habit of growing in one or two generations.

Thinking about fire as trees experience it enables us to understand its very essence more. When we watch a flame licking along a log, what we are seeing is the outline of a bubble of gas—pyrolyzed wood, the part of the log that has vaporized in the heat—reacting at a molecule-thin interface with oxygen and giving out energy as light and heat. If you burn apple-wood on your fire on an autumn evening, the blue at the edge will be the sweet-smelling alcohol that gives off a faint scent of apple, but most of the gas will be made by the pyrolysis of cellulose and lignin, happening around 300°F and 572°F, respectively. The flame is delimiting the edge of the bubble of little molecules that splits off from these big molecules as they heat up; first the oxygen- or hydrogen-containing molecules, then more and more pure carbon chain, burning hotter and hotter. Charcoal burners exploit this process by burning wood slowly in a limited air supply, causing all the oxygen- and hydrogen-containing molecules to burn away until only pure carbon chains are left, black and solid.

For a tree, loss of carbon is the loss of the structure of life itself. It may not always kill the tree, but it can wreck the structure and integrity of its trunk, the scaffolding by which trees reach up to sunlight, which in turn leads to it being shaded out. A living tree caught in a forest fire will experience the smooth initial heat of its alcohols burning outside the bark, then the drying effect of water being sucked out, and then, as the fire bites through the bark, the vaporization of its very structure; a painful eating into the heartwood as heat increases and lignin pyrolyzes. Alcohol vaporizes quickly, and du Châtelet's argument that "spirit of wine"—brandy—burns more readily and at lower temperatures because it separates out and leaps toward the flame is exactly right. If trees produce alcohols that are highly flammable but burn cool, they can use fire as a form of protection—a mock fire at a low temperature to prevent a real, hot fire from happening.

There's a game called snapdragon that my sister and I used to play, cross-legged on the hearthrug during the long, wet Devon winters. Dickens refers to it, but the best evocation of it comes in John Dryden's seventeenth-century drama *The Duke of Guise*. "I swallow oaths as easy as snap-dragon," says the conniving curate, "Mock-fire that never burns." The game requires a shallow dish filled with brandy and raisins, which is warmed slightly and then set on fire. You snatch out the burning raisins, eating them while they're still alight, and although the blue flame will run down your hand and you will sometimes scorch your fingers, normally you can carry the mock fire to your mouth and snap your jaw shut on the burning raisin unscathed, feeling only an otherworldly tingle of the air on your arm. This is exactly what some trees, by the uncanny designs of evolution, have elected to do.

Trees have been playing a form of snapdragon for 309 million years at least, since the late Carboniferous scale-trees, *Lepidodendron* (from *lepido*—scale, and *dendron*—tree), began to encourage a sort of mock fire that would whisk through the understory and leave them unscathed.★ Standing underneath a scale-tree grove you'd find yourself in a deep leaf litter concealing a treacherous, oily swamp. On the trunks massed close around you, shiny green diamond scales would twist into helical tessellated columns like a baroque church in Munich. These trunks, completely bare and shiny for the first 16 feet, would then be clothed in leaves like pine needles. Long needles from

---

★ There is evidence for fire among species in the Silurian and the Devonian, but the charcoal deposits are not extensive, and there is no good evidence that trees adapted to overcome it at this stage. See Scott, A. C. (2000). "Pre-Quaternary history of fire." *Paleogeography, Paleoclimatology, Paleoecology* 164, 297–345.

the center of each scale would overlap all the way up the trunk to the strange, fractal canopy. As the tree grew, its needle-shaped leaves would drop from the scales of the trunk to form an aerated, bouncy layer full of flammable alcohols. If a fire came, it was likely to whisk through the leaf litter, destroying predators and liberating nutrients, but unlikely to seriously burn the trees.

Mock fire in the *Lepidodendron* leaf litter was a protection against serious fire. Scale-trees grew as tall as 100 feet and could live for as long as a hundred years, so needed a way to protect their longevity or at any rate prepare the ground for their children. Mock fire would often have raced fast through the aerated leaf litter and burned cool, unlikely to cause big conflagrations, and therefore incapable of scaling those smooth green columns, incapable of producing large cinders that could wedge themselves into the canopy and cause serious fires. The scales, like the bark, were fire-resistant, and the trees all the same height, with no other species to act as ladder trees for the fire to reach the canopy.

Fire resistance was necessary. Fire follows fuel, and trees contain all the vast energy of the sunlight they have captured over their lifetimes, locked down in the bonded carbon of their structure—the perfect fuel. A crown fire, in which the tops of the trees kindle, is an awesome sight, sucking up air from the ground and jumping from tree to tree. In its most dramatic manifestations it can create such an enormous updraft that it forms a tornado. But when did trees start to shape fire? When did they start to adapt their shapes and the very fabric of their being so they could exploit it?

The record of fusain—ancient, fossilized charcoal—shows that wildfires started to spread almost 383 million years ago. These earliest fires followed the expansion pattern of the lignophytes,[4] trees such as *Archaeopteris* that were some of the earliest true trees,* growing more than 330 feet tall by hardening their veins with lignin and inadvertently becoming tall chimneys of fuel as a result. It may even be that fire was the reason for the extinction of *Archaeopteris* at the

---

* They thrived by using extensive root systems to explore drier sites like the Cairo site in upstate New York mentioned in Chapter 1.

end of the Devonian period, a mystery that has concerned dendrologists for years.

Archaeopteris grew on drier ground than the trees that survived, the
forerunners of *Lepidodendron*. Its leaves resemble ginkgo, a tree so fire
retardant that Buddhist monks planted it around shrines to protect
them from burning, but oxygen levels were higher by the end of the
Devonian than they are now. What if fires smoldered and raged so
badly and constantly through the upland forests of *Archaeopteris* that
the whole species was fragmented and the population collapsed, leaving only trees of different species that were wildfire-adapted or lived
in swamps? It could even be that wildfire through the dry *Archaeopteris* forests was a major, if not the only, reason for the extinctions
at the Devonian-Permian boundary: three great pulses of extinction
that killed over 90 percent of the corals and marine life that lived in
shallow seas. There's good evidence that a peak of phosphate and
nitrogen in the sea caused algal blooms and subsequent anoxia and
extinction, just like cases of eutrophication in which rivers, lakes,
and seas die when fertilizer or sewage is released untreated. At this
point only tree roots had the capacity to mine the fresh bedrock so efficiently that this type of overfertilization, in the form of phosphate
and nitrogen, could be released from rock. What no one knows is
how exactly this phosphate and nitrogen were washed into the sea.
However, fierce fires in the huge upland forests of *Archaeopteris* could
have produced smoke particles that provoked deluges, sweeping the
phosphate and nitrogen of the ash into the water.*

Many branches of the trees' family tree may have been burned off
by fire at this early stage, leaving the fire-resistant trees and making
fire resistance or adaptation increasingly important. Trees like *Lepidodendron* boomed in the late Carboniferous period, oxygen levels rose,
and fire became even more rampant.[†] There were several reasons:

---

* This could have had just as drastic an effect as the other postulated drivers of
extinction, which include: global cooling as a result of a surge in oxygen, a mega
asteroid hitting earth, or a reduction of ozone leading to terrible UV-prompted
mutations.
[†] By the late Carboniferous wildfires were extremely common, more common than
at any time apart from a peak in the early Permian, 290 million years ago.

with oxygen comprising 32 percent of the atmosphere everything burns, the temperature was approximately 11 degrees higher, and, most importantly, fuel, combustible matter, was everywhere: cala-mites like huge horsetails, scale-trees, and the conifer-like cordites. *Lepidodendron*, perhaps descended from trees that had been protected from extinction by growing in swamps too wet for fire, were at huge risk in this oxygen-rich environment, but because fire is dependent on very specific local conditions, each tree, if it grows with trees of the same species, has control over which fuel might burn it.

If they failed to survive using mock fire, scale-trees had another trick up their sleeve. Scale-tree groves, like a rugby scrum in their monocultural magnitude, were probably much stranger than they appear. Their cones developed precisely on the growing tip of the plant, so once a plant became fertile it could no longer grow larger. A grove of scale-trees would all grow together, therefore, drop seeds together, and die together. Dying all at once is an advantage in the case of fire. The light, spongy wood, held together by an exoskeleton of harder scales, can burn fast; assuming the seeds were fire resistant or safely insulated by a layer of unburnt bog held together by their parents' roots, they could all have sprung up together, the next co-hort rising together from the fertilized bog.

Lepidodendron are now extinct, but trees still use some of the tactics scale-trees developed to survive fire, and from some of them we can learn more about how *Lepidodendron* coped with fire and when it used mock fire than from the fossil record alone. Today, for example, almost a hundred species of trees, some very rare, use the strategy of reproducing all at once and then dying. Often this is triggered by damage, of the sort that might be caused by a fire or a drought. One example is the rare candelabra tree that grows in New Cale-donia, *Cerberiopsis candelabra*, which tends to flower in years when it has experienced a tropical cyclone—as does the Brazilian suicide tree, *Tachigali versicolor*—or outbreaks of lightning or strong winds.[5] Other trees still use mock fire—a cool, light, sweeping fire—to protect themselves from hot fires that might annihilate them. Pines, as well as using fire as a weapon against competition from fir trees that would outgrow them, use mock fire as a way of killing herbivores and pathogens and releasing nutrients. One pine in particular, the chir pine, *Pinus roxburghii*, has needles of exactly the same dimensions as some *Lepidodendron* leaves, and its cycle of mock fire and regrowth probably shares many similarities with this very different tree that flourished 318 million years ago.

The chir pine clothes the slopes of the Himalayas and the Hindu Kush from Afghanistan across Pakistan, India, Nepal, and Bhutan to Myanmar, and its cycle of mock fire and regrowth is intrinsic to the tree's life. Imagine you are walking 6,600 feet up in the foothills of the Himalayas near Biratnagar in Nepal. The pines are tall and close together, and very few plants can grow underneath them—occasionally a scrubby banj oak or some heath-like shrubs. Otherwise, there is a clean floor of brown needles, long and held together in bunches of three, hardly packed down at all, bouncy and dry, so that with every step you take up the hillside you slide down a little. The smell of the forest is of pine: not the fresh, wet smell of a pine forest in Finland, but a clove-scented, black-pepper smell, with a background of turpentine from an oil-painting workshop, and top notes heady with a rush of aromatics. Some of the volatile organic compounds the chir pine releases will evaporate upward and help to seed clouds, but others, not so volatile, will hang around. If you were to lie on this pine needle mattress the smell would be stronger still, mainly because of terpinen-4-ol,[6] which makes up 30 percent of the liquid in pine leaves, while 32 percent is (E)-caryophyllene, the fresh smell of crushed black pepper, but also small amounts of other alcohols like hexanol.

How deliberately a tree shapes this mixture of volatile compounds to encourage a fire to burn to its advantage is impossible to say, but we know that the scalpel of centuries has carved out seeds that scarify to germinate only at certain temperatures. It wouldn't be too much of a stretch, therefore, to accuse the Magimix of millennia of concocting the perfect flaming mock shot. The miasma of alcohols hovers just above the pine needles, mixing with oxygen, and were a spark to come and heat the surroundings to just 174°F (the flash point of terpinen-4-ol) or perhaps even lower,⋆ a reaction zone would sweep across the ground and a ripple of flame would be pulled in all directions. The flame would heat up and penetrate the air pockets in and around the needle-mattress, all along the surface of the needles, until the lignin and cellulose of the needle catches and produces a visible yellow flame, along with a thin line of smoke from the longer carbon chains that haven't burned cleanly. Beetles have little time to escape from the swift-moving fire; they're asphyxiated and their wings immolated. Fungus too is unlikely to survive. A diet of needles alone, glowing red and carbonizing in a puff of smoke, will be eaten through fast, and the fire will quickly die down from lack of fuel.

Thus far the pine trees have endured a mock fire they can probably survive; if some needles are browned they will quickly green up again. However, when fire reaches the chir pine's trunk a negotiation ensues. Fire is essentially seeking out volatile carbon compounds, sniffing them out as the next probable place for a reaction to occur. Here, there is little that is immediately volatile: terpinen-4-ol is only 4 percent of the bark's makeup, so if the fire is not too hot it will make short work of the pine needles. But (E)-caryophyllene makes up 35 percent of the bark's mass, and if the temperature rises, the flame leaps high enough, or the fire finds enough fuel for the bark to heat up to around 248°F, then the caryophyllene will vaporize and the bark too will start to burn.

Cones of the chir pine take twenty-four months to grow and mature, turning from photosynthetic green to brown and opening slowly over the second year. That's unless a forest fire pops them open, releasing the seeds all at once. This doesn't happen by coincidence—there's a molecular mechanism involved, one of many, which, when advantageous to the pines, will flip a switch and turn mock fire into real fire, a slow-burning, hot fire that clears everything and allows the pines to start again from seed. Where a tree has already

---

⋆ Compounds present in chir pine needles such as hexanol or alpha-pinene have even lower flash points, of around 111°F. These molecules are only present in small amounts, although their production is often upregulated in drought.

been damaged by fire is one such situation, and here we may find *jhukti** on the ground, brilliantly colored heartwood imbued with a crystalline and brittle resin of red or yellow that will ignite easily and burn long, a substance traditionally used for lighting and starting fires.[7] It's a kind of suicide note, to encourage hot and lingering fires to burn the old tree and allow new young pines to flourish.

Another scenario is where a pine faces serious competition, and there are other fuels to keep the fire going. In the case of the chir pine, rhododendrons, the banj oak (*Quercus leucotricophera*), and various types of heather—can all grow under it, and in this case the bark is more likely to burn through, and the fire that licks up into the heartwood of the tree can carry on burning for some days, inducing the loss of volatile substances and the formation of *jhukti* if the tree itself does not burn completely.

Other scenarios include drought, when wood is more likely to burn hot, or when there has been no fire for some years. In this case beetle-bored branches and dead trees (*jhukti* resin will appear in response to mechanical damage as well as fire damage), cones, and deep-decayed needle litter will offer the beginnings of a grand conflagration—not a mock fire but a true wildfire. A whole variety of scrubby, deciduous oaks, the banj oak among them, which contain fire-retardant chemicals called tannins, will also sprout again quickly after fire; it's what allows them to compete with the chir pine's mock fire approach. But if a true conflagration does break out, the chir pine seeds, freshly dropped, will germinate within weeks, whereas the banj oak will struggle to recoup its strength and the carbon evaporated by the fire.

Lightning, whether in or out of the hands of gods, is the most blatant manifestation of the extreme power of nature—a terrifying, ex-

---

* *Jhukti* is similar to the wood used by the ancient Maya for their ritual sacrifices in what is now Belize and Guatemala. Wood from the Mountain Pine Ridge in Caypo District has been found in caves 50 miles away, probably because the resinous wood burned for a long time and gave out a good light.

traordinary force. When a storm cloud forms, water droplets in the cloud fall, only to be swirled upward in vortexes of wind and break up again, and as this happens over and over again the highest part of the cloud becomes positively charged, the lowest negatively.[8] As the cloud moves across the surface of the earth, the earth reacts to mirror it and, like a photographic negative, electrons are pushed down away from the clouds, leaving exposed points—rocks, chimneys, the growing tips of trees—vulnerable and positively charged. In the moment of a lightning strike, for some few millionths of a second, electricity pours through the tree and the heat vaporizes water, sap, and the wood itself. I once saw a splinter of trunk 50 feet long and tons in weight impaled in the ground as though a giant had been throwing the javelin.

Around the Pacific, a ring of fire is maintained not just by volcanoes but by conifers that stand like ziggurats, layered temples to the lightning gods. The kauri trees of New Zealand, the redwoods of California, the *Fitzroya* of Chile, and the cedars of Japan grow to great ages and great heights, trees of huge charisma that tower above everything. Ironically, however, all depend on catastrophic fire to regenerate. More widely spaced than most trees, after they have been growing for about a hundred years many of them will change their whole conformation, their canopies flattening out. But until then they are like enormous matches, an irresistible route for lightning out of the sky and into the earth.

Similar lightning conifers once created a belt of fire around the earth so impregnable that for 50 million years they prevented dinosaurs from dominating the world. We know this from rainbow fossils, trees alchemically preserved in the Arizona desert. *Araucarioxylon arizonicum* is the state fossil of Arizona[9] and the resinous forerunner of the monkey-puzzle, the rare and beautiful *Araucaria angustifolia* of Brazil, the kauri of New Zealand, and the metal-twisting *Araucaria* species of New Caledonia. In its fossilized life the tree is crystal, the wood permeated by minerals that dissolved out of volcanic ash and transformed into solid quartz, colored by iron and manganese and carbon to sparkle with citrine and amethyst and jasper. In life, 220 million years ago, the trees could reach up to 200 feet tall and had a tapering trunk with a diameter of nearly 2 feet, supported by a taproot that ran down 16 feet and branched only occasionally. All the various parts of the fossil tree that have been unearthed have the characteristics of a fire-promoting yet fire-sensitive tree. Most of the bark is lost from the fossils, but the few samples that remain are like the bark of Aleppo pine or the chir pine, thin and ridged with long furrows.★

No one knows exactly how often fire swirled round the ecosystem these ancient trees inhabited, or how they recovered from it, but the silicate crystals allow us to analyze fire scars preserved within the wood. They show patterns compatible with mock fire: a period of smaller tracheids and lower growth consistent with drought is followed by a burn on the leeward side of the tree, characteristic of a low-intensity undergrowth fire. After a time of recovery, in which the tracheids are again small (and sometimes malformed), comes a season of magnificent growth and health, in which phosphorus and nitrogen from ash flood the soil and reduced competition allows the tree to flourish phoenix-like from the ashes.

In the brick-red rock of Ghost Ranch in New Mexico, paleontologists have found proof of much more serious wildfires. These burned at up to 1256°F across the unstable conifer forests of the equator, forming a wall of fire that altered the ecology of the world for millions of years. It was this belt of wildfires that prevented the big dinosaurs moving into what is now North America, keeping them pent up in the south, and it was only when this fire belt diminished, at what we now call the Triassic-Jurassic border, that the age of dinosaurs could truly begin.

Fossils in those rocks show that about 210 million years ago Ghost Ranch was at a latitude equal to Kerala in southern India today, and the mean temperature was about 82°F. During a similar period to the one in which the huge vegetarian dinosaurs, forerunners of diplodocus, thrived elsewhere,[10] raging wildfires and great boom-and-bust cycles

---

★ More research has questioned whether *Araucarioxylon* actually exists, or is simply a chimera—the remains of many species squished together like a Frankenstein monster to create an impossible tree that never existed. https://www.sciencedirect .com/science/article/pii/S1631068310001260

in vegetation were common. In wet periods tree-ferns flourished, but drier periods saw conifers such as pines thrive. Intense wildfires then swept through the pines, making it impossible for large plant-eating dinosaurs to survive.* Perhaps it was better for the trees to burn and recycle their nutrients rather than be grazed on by huge, rapacious herbivores that could carry the nutrients they contained over the nearest hill. Certainly, the high carbon dioxide levels of the atmosphere made vegetation run riot, while the high oxygen levels and high temperatures made it likely to burn. Trees had to get fire on their side.

Over the next 200 million years, continental drift swept the rocks of New Mexico northward, drought obliterated the forests, and wind and water cut swathes through the red rock. When the artist Georgia O'Keeffe came to settle in Ghost Ranch, she found dramatic slices of time preserved in the fossils of the stratified rocks. "The first year I was out here," she wrote, "I began picking up bones because there were no flowers."[11] Subsequent researchers have described the bones of small carnivorous dinosaurs "practically oozing out of the sandy red soil." O'Keeffe played with the sense of antiquity laid bare, and in some of her paintings the dead are more present than the living. In *Gerald's Tree*, the bleached skeleton of a tree stands against the backdrop of red sandstone.

The fire belt of the Triassic fragmented as the continents moved apart, but amazingly its after-effects are still with us today in the great resinous conifer stands that have survived around the Pacific. The continent of Gondwana moved south, and among its remnants are the island of New Caledonia, with its superb and strange metal-eating araucarias, the alerce forests of Chile, and the Wollemi pine enclaves of Australia. The surfboard of the kauri trees was Zealandia, and the islands of north and south New Zealand are its non-submerged

---

* It is probably because of this that paleontologists haven't found a single (plant-eating) dinosaur near the equator during the Triassic, whereas thousands of examples have been found in South America, southern Africa, and Germany.

highlands. While most trees in New Zealand are not fire adapted—in the cool, wet conditions they don't need to be—kauri trees, which dominate the forests north of latitude 38°S, use lightning and fire as part of their world, just like their forerunners crystallized into quartz in Arizona. To our eyes kauri look like pillars of peace, the torque on their trunks an evocation of eternity, but they use fire to create chaos. Mapping their movements in New Zealand over time shows them retreating uphill during periods of calm, then taking over lower areas briefly during the mass disturbances of flood, fire, or human interference. It seems a strange tactic for a tree that can live to 2,000 years, but kauri encourage and promote fire as part of their ultimate survival and, as trees did 200 million years ago, they do it by means of a chemical resin.

To the nineteenth-century colonizers of New Zealand the most famous of the kauri trees was called the Father of the Forest, a tree 72 feet in girth. In 1890 it was struck by lightning and died, but eyewitnesses describe how it was found packed with huge quantities of resin, which had flooded into the tree in response to the lightning strike.[12] Resin is made up of two fractions: the volatile short chains that allow the resin to flow in the tree and evaporate off as the resin hardens in the air, and the longer chains that align to form hard, brittle, clear resin (and later amber). For those European colonizers kauri resin was hidden gold: between 1850 and 1950 half a million tons was exported from Auckland to be used as an incredibly resilient varnish for violins and the principal ingredient of paint.[13] The Māori used it as chewing gum and as a fire starter. For trees it can seal wounds and deter herbivores, but for wildfire it is a time bomb.

This is more than mock fire. Just like the *jhukti* resin of the chir pine used as a firelighter in Uttarakhand or the pine resin and torchwood used to illuminate underground cave rituals of the ancient Maya, kauri resin can be easily ignited and smolder for a long time. When a tree is struck by lightning and fills with resin, the next lightning strike will blow it up into a serious conflagration, burning very hot.* Resinous

---

*Mapping fire and trees and the evolution of both across millions of years of changes in oxygen concentrations and heat is like juggling hot plates on a roller-

damaged trees would burn intensely; undamaged trees would not. A wounded tree can smolder for days or months, and if there are sufficient numbers of dead or dying trees nearby it can spark off an epic fire capable of regenerating half a forest. Once upon a time, perhaps in the age of giant beetles and huge caterpillars, insects were the target of this sacrificial deep cleanse, but resin is also effective against the diseases that strike the kauris nowadays, like the oomycete kauri dieback, a *Phytophthora* that is killing reserve after reserve of kauris in New Zealand. In response to the fungus-like oomycetes, kauris produce copious resins, sometimes ringing the tree with it high up in the canopy. Then a lightning strike! And the resin that prevented the insects and the oomycete from gaining too much of a foothold will kindle a fire of sufficient intensity either to sweep the threat away or keep it in check, hopefully for another 2,000 years.

In Māori myth, the resin of kauri trees is what is left after the tree gave its oil to make blubber for the southern right whale, as its insulation from the cold waters of the far south. In the myth, the whale in return gave the kauri tree its scaly flaking skin, to protect it from epiphytes,* the implication being that epiphytes could grow so heavily on an old tree that they could actually pull it down. Look at an old kauri and you can see where the idea comes from. For the massive old conifers of the fire belt and their descendants on Gondwana, fire was an inescapable reality that had to be turned to their advantage if they were to grow in the wide-spaced forests they needed. As the continents drifted apart, a new and faster world of non-coniferous trees developed. Most old conifers are now waning, unable to keep up.[14] Some of the original conifers, like the cypresses, evolved away from producing resin; others, like the pines, diversified into different

---

coaster; accounting for the interplay of ecology under the changing conditions of geological time is akin to operating an Astrojax in space, but by using amber records from around the world, as well as the similar chemical composition of resins (abietic acid, agathic acid, and sandaracopimaric acid come together to produce the clear substance), we can piece together a pattern.

*Neither species, of course, had a defense against humans who have been their latest bane, destroying 90 percent of the forest between their first settlement and 1900.

modes of living fast and dying young, and the new trees found different ways of shaping fire.[15]

Pyrogenic or fire-promoting trees grow all around the world and tend to develop fire regimes: patterns of fire or mock fire that are quite predictable. Consequently, the trees they grow with will work around the fire or mock fire they employ, finding ways to avoid or control it. This is what has happened on the island of Evia, a tree heaven, where an oak tree that would normally burn down and not resprout has rapidly evolved to live with fire.

Evia is a mountainous island along the east coast of Greece. It is full of different habitats, all occupied by different trees, but one of its rarest lives in the north, an oak tree that flourishes among the pines and is only known to grow on this one island: the Evian oak, *Quercus euboica*. I had been to Evia a few times before: once by the sea among the pines at Limnionas; once climbing in the mountains of the south where junipers clung to thin soil; and once to see the sweet chestnut forests flower in undulating white clouds over the hills in the middle of the island. In August 2021 fierce fires drove villagers into the sea and destroyed 141,000 acres of forest, and in May 2023, nineteen months later, I went to see if any of the *Quercus euboica* population had survived.

I stayed with two Greek friends who farm walnuts in northern Evia, and we drove through the area of the fires to where the oaks were recorded. The pines made a sad black hedgehog of the mountains, the carbon they had carefully sequestered over many years returned to the atmosphere. Thin charred sticks dangling their blackened and burnt cones like little bells were all that was left of the canopy, but underneath was a mass of green and yellow and pink as the arbutus sprang back to life, the convolvulus grew rampant, and the cistus flowered like fury. Every so often there was a patch of unburnt trees or a verdant and apparently untouched meadow bursting with flowers, but mostly it was ridge beyond ridge of burnt mountains.

These blackened sticks had once been fresh green pine forests of the Aleppo pine, *Pinus halepensis*, one of the most pyrogenic trees in the world. In spite of its name it is native to Italy, Spain, France, and Greece, making up 8 percent of the forests there but causing one-third of all the forest fires. The only reason we know it as the Aleppo pine is because it was named in 1768 by the botanist Philip Miller, based at the Chelsea Physic Garden in London, who had probably been sent the seed by the consul in Aleppo.[16] The pine can survive severe drought, has very thin bark, and is killed by all but the mildest fires. But it regrows extremely fast from the serotinous seed that drops out of the cones that remain on the burnt trees, so it tends to set up a fire cycle on a 15- to 30-year basis.

Growing among the Aleppo pines, oak trees had to adapt fast to fire. Evian oak is a subspecies, related most closely to the Trojan oak that grows on the plain surrounding the ancient city of Troy in Turkey, 200 miles away on the other side of the Aegean. On the plain the Trojan oak is surrounded by grass and grows directly up into a tall tree with a single trunk; fire will kill it outright and it will not regrow. In the mountains, surrounded by pine, the Evian oak, on the other hand, thrives on fire and quickly resprouts. My friends led me through the burnt trees up a red track onto the ridge. Here, all was still black and sooty, the blackened bark flaking off the pines to reveal a white, dry wood. Pine saplings were already coming up, but the real surprise was the little copse of oak on either side of the road. Burned to the ground, it had sprouted again in thick glossy shoots, full of pigments that gave a red tinge to the first leaves, rising out of the ashes faster than any tree could grow from seed.

Quercus euboica grows on serpentine, a soil full of iron and magnesium that give it flaky red and purple layers, which contains very little nitrogen. As the fire whirls through the pines and burns the upper levels of the oak and the vegetation that protects it, nitrogen and phosphate break out of the organic matter that held them and fall to the ground in the ash. The soil's fantastic insulation properties will protect the meristem, the underground growing point of the tree, from the heat. Once it has been enriched by ash, the soil will contain accessible nitrogen and phosphate, and the oak trees will grow up fast from the burnt ground younger and more vigorous than before. The custodian of the trees told me they had regrown within two months of being burned.[17] The Aleppo pine, on the other hand, the stoker of the mock fire, will take two years to grow up beyond the oaks.

Sometimes fire doesn't work for pines—in the Sierra de Alcaraz in southeastern Spain, Aleppo pine is often outgrown after fire by shrubs like the prickly oak, *Quercus coccifera*, *Cistus*, and the grass *Brachypodium retusum*, which force the pine to the even greater aridity of south-facing slopes. In these Evian mountains, too, the cistus was thick on the ground. Across the Mediterranean this deceptively simple and fragile shrub reigns supreme, a brittle-branched shrub with flaky brown bark, dry sagebrush leaves, and simple white or pink flowers like a Tudor rose with five crumpled petals and a simple yellow heart to entice the bees. Its success is down to a tolerance for drought and a superb fire adaptation. Cistus produces two types of seed: one can germinate easily, whenever there is enough moisture, but the other requires scarification by fire before it can do so. This means that while most of the seeds come up and flower year after year among the competition, a proportion can lie hidden in the soil and sprout twenty years later when a fierce fire has swept the earth clear. It's very likely that other species of tree will prove to have this brilliant mechanism of fire adaptation, but no one has described it yet.*

A big fire on the mountaintops is, as Homer writes, something that "blazes through the sky to heaven, an amazing sight." Whole mountains become flame, fire can leap across ravines, heat can advance across rivers, and yet, in the midst of all of this, one small patch can be burned around and remain almost unscathed. Normally this is the result of a different tree species shaping the fire. Just like the firs on Mount Kazdaği, many other trees have adapted to repress fire. On northern Evia, the beautiful walnut groves by the coast at Kefales stopped the fire to within a few yards, and their mode of doing so is unique to the Juglandaceae family, whether in America, Europe, or

---

* One likely candidate is the crown-of-thorns tree, *Paliurus spina-christi*, so called because it was meant to be the tree that made Jesus' crown of thorns on the cross. Just like cistus, its seed will germinate only when it has been scarified, but this could also be an adaption to wind dispersal.

Asia. The key is simple and brilliant: a chemical, juglone, which is not only directly fire resistant but also poisons nearby plants that might act as ladder fuel and help the fire up into the leaves of the walnut's canopy. Juglone is a deceptively simple chemical, a flat double ring of carbon that is stored in walnuts in its yellow, hydroxy form and then oxidizes in contact with air to form the black compound. Roots, leaves, husks, bark, and fruit (although not the nut itself) all contain high quantities of juglone, which is still used as a dye by American Indigenous peoples, while tapestry weavers in Central Asia and the hairdressers of ancient Greece also valued its color-fast glossy brown.

As a dye, the planar rings of juglone slot inside helices of keratin and reflect only the yellow-red-brown shades of light, in a similar way to its chemical isomer henna. As a poison, juglone slots itself into enzymes required for metabolic function, blocking the electron transport chain in mitochondria and causing electrons to pile up and make reactive oxygen species, which then cause havoc. The same mitochondrial action makes it lethal if eaten by most insects, and when transmitted through the air juglone can prevent stomata from functioning. Interestingly, plants in its own family, including birch and beech, are far less likely to be affected, which suggests that walnut developed this poison early on in its evolution. In common with oak it appears to have developed tannins at an early stage, which are effective as an insect deterrent and a fire retardant. Walnuts have thick bark, but rely more on their heartwood and rot resistance to protect them from serious fires, and if burned to the ground they will shoot up from the base.

How is it that chemicals such as tannins and juglones act to repel insects as well as suppress fire? Evolution is ruthlessly efficient, and so the same chemicals that afford this dual defense evolve time and time again in different tree lineages as they fight to adapt to evolving threats. The same is true of physical adaptations: although the cork oak, *Quercus suber*, contains tannins, it places almost all its faith in its thick bark, a physical adaptation that can be just as effective as a chemical one against mock fire. The tree is not very combustible, but the plants that surround it tend to be, and it is cork oak bark alone that means an ordinary fire will leave the oak unscathed. A severe fire will burn the foliage and twigs, leaving trunk and branches that resprout, and a repeatedly burned cork oak may be deeply scarred, but is rarely killed outright. An adaptation like this in a slow-reproducing tree implies hundreds of thousands of years' exposure to fire and is almost unique. Cork oak produces the cork of wine bottles, spongy dead cells filled with air whose thin walls are made of cellulose and the waterproof material suberin, a lipophilic polymer that acts like a liquid plaster to permeate and lock off cells. Cork is produced by the phellogen layer of the tree, and its harvesting as cork requires multiple stages. First the useless cork is stripped off—this will be brittle and cannot be used for anything. The initial regrowth can be used for flooring or shoes, and then in the third or fourth stripping the cork is ready to use: spongy, a consistent texture and clean enough to plug a bottle of Riesling without altering the flavor.

Humans, like trees, tend to discover by trial and error a patchwork of productive methods of setting a small fire. If they burn wood, they start to stitch together a hierarchy of burnability, an innate knowledge of the chemical and mechanical structures of wood—thin bark, resin, hollow stalks—which make it good or bad for burning. In *Homage to Catalonia*, George Orwell talks about the endless hunt for fuel to warm him and his companions as they shivered in a dugout in the depths of winter:

> The eagerness of our search for firewood turned us all into botanists. We classified according to their burning qualities every plant that grew on the mountain-side; the various heaths and grasses that were good to start a fire with but burnt out in a few minutes, the wild rosemary and the tiny whin bushes that would burn when the fire was well alight, the stunted oak tree, smaller than a gooseberry bush, that was practically unburnable.[18]

The oak in question is probably the cork oak, and the chemicals that made it practically "unburnable" are suberin, tannins, and lignin, a fire-retardant mix refined over millions of years.

Over much greater areas and longer timescales humans have often interfered with the ways that trees shape fire, and because of their imperfect understanding of both trees and fire this can be a disaster. Fire suppression was a forestry policy instigated by those (normally trained in the traditions of German forestry) who were looking for timber that grew straight and clean.

One place where fire suppression was strictly enforced was California, where the great redwoods, *Sequoiadendron giganteum*, have used mock fire to regenerate for 200 million years, all through the Jurassic and Cretaceous periods. They did this by producing resins and highly flammable needle drop and by only releasing seeds from the cones they retain on their branches when the heat of the fire dries them out. They also developed spectacular fire resistance. Large trees can have bark 3 feet thick, hairy like asbestos, which shrivels and retreats without fueling the flame. When that is breached the wood weeps bitter pale-pink tannins, repelling insects and fire together. The result used to be a regime of frequent low-level fires, burning small patches of vegetation along the ground and rarely becoming a crown fire, the roaring, raging updraft fire that can set the canopy of a fully grown tree alight.

In the past 200 years low-level fires have been suppressed. When Western colonists reached the redwood forests of California they marveled at their beauty and extraordinary size—and then destroyed so many of them that finally there was a public backlash and a conservation movement was formed.★ In the latter half of the nineteenth century, Americans, particularly those in New York, were alarmed by newspaper accounts of a succession of conflagrations that had burned millions of acres in the upper Midwest. The fire-suppression service that resulted became one of the best trained and most effective in the world: Cal Fire. Cal Fire stopped Indigenous peoples from lighting fires and had a zero-tolerance approach to fire that later became its "10 a.m. policy": any fire spotted would be stamped out by 10 a.m. the next day. George Stewart's novel *Fire* brilliantly evokes the ethos of the firefighters of this time: macho, well funded, and organized

★ The complete story is brilliantly told in Lyndsie Bourgon's book *Tree Thieves*.

like a pseudo-military operation, with fire spotters on perches looking over the forest, planes, and bulldozers.

Unfortunately, suppressing small fires allows fuel to build up, and often the result can be cataclysmic. After a series of big fires, Cal Fire started to investigate the possibility that the Indigenous peoples had in fact had the right idea when they set small fires to prevent big ones and has now itself adopted the practice of controlled burns. The question is, is it too little, too late? I went to California to marvel at the redwoods, but also to visit Cal Fire and watch the fire service conducting a controlled burn—essentially to see what it looks like when humans use mock fire.

Up in the mountains among a group of famously big sequoias, the Trail of a Hundred Giants, the fire service had already started its controlled burn. As I drove through the trees toward the parking lot, I saw sunlit smoke trickling through the trees and little tongues of flame licking up from piles of brash. Watching from a safe distance I was struck by the confusion and the noise: a controlled burn is not as simple or as gentle as it sounds. Pitch-black smoke was in the air, and the smell of fire was acrid and ubiquitous. Orange tongues of flame raced up the hill, round the trees standing and lingering on their upper side, and stopping only when they reached the dead areas called backburn, the scene of a previous fire. Every so often there was a bang like an explosion, as a massive tree was cut down to protect the sequoias. The felled trees were giants in their own right, mostly great pines with lacework bark.

Ignacio was a member of the federal fire service responsible for the controlled burn. He knew fire, had grown up in an area of Mexico where forest fires were common. A fire fighter since 2001, he now drove thousands of miles around the country fighting for the safety of houses and forests. A bad fire soon after he arrived with the California service made many of the service quit—they had been running for their lives, eyebrows singed, chased by fireballs. The controlled burn program here was being enforced, Ignacio explained, because the intensity of the fires, which had in the past rarely harmed fully grown redwoods, was increasing to the point where the year before a fire had got into the grove. The huge fire had begun near Black Mountain on the eastern side of the Sierra Nevada and swept across it, imperiling the Tule River Tribe reservation and tickling the edge of the redwood grove.[19] Luckily, it only slightly singed most of the redwoods in the grove, but it was a close-run thing.

As a result, the federal fire service and Cal Fire had started a joint program of controlled burns to reduce the fuel around the site. They were using the pile-burn method of prescribed burning: felling huge pines and small cedars and burning them along with the undergrowth. Most giant sequoias will not have any branches near the ground, so fire is unlikely to reach them unless ladder fuel—smaller pines or other trees nearby—allows their leaves to catch fire or the fire moves through their bark and into their trunk. If this happens they have a second line of defense, chemical tannins that suppress fire, but once their bark is no longer there to protect them they really burn. A cinder like a fist will get caught in the bark where nascent branches have fallen off, and then fire will race down inside the trunk, smoldering and burning for days. Red cedars are the least flammable of all the trees in this area, possibly due to the long-chain fatty acids they share with other cedars worldwide, all strong fire suppressants. Sequoias, too, are naturally fire resistant, but can be overwhelmed by fires that have gathered speed and become extremely hot. In the past, trees with trunks over 13 feet thick were rarely affected by the fires, but in the past ten years more than forty of these trees have been killed. Fires are getting too large and too hot to escape.

In places where controlled burns had already happened, burn marks showed up to 65 feet high on the redwood trunks—marks they would carry for the rest of their lives. Even with the immense fire trucks of Cal Fire and the well-organized controlled burns, fire is an existential threat for the giant sequoias. Over the past few years it has wiped out 20 percent of the largest trees, and most of the trees in the grove have signs of former fires on them—great gouges out of their bark, hollows, burn marks around their bases, and curious flame-shaped arches, called cat-face arches, where the fire has stuck. These fire scars embedded in the trunk are one of the reasons we

know so much about the intensity and change of fire regimes in California.* A slice across a redwood trunk may give you a chronology
of fires and droughts stretching back 3,000 years.[20]

Professional foresters love to plant fire-promoting trees like pine, because they are trees that have evolved to grow fast and straight and
happily in a monoculture.[21] Frequently, however, this is counter-
productive and results in terrible fires. Worldwide, the outstanding
example of this is the eucalyptus.[22] Flaky barked, silver leaved, fresh
scented, and koala associated,[23] for me it is a terrifying and toxic tree,
the only one I might actively dislike, for it is a tree perfectly adapted
to work with fire, to the point of positive detriment to the surrounding species. Extremely fast-growing—it can grow up to 11 feet a
year—it has been planted for forestry all over the world, covering
more than 77,000 square miles in more than ninety countries.[24]

It is a catastrophic choice, particularly if planted, as alas all too
often, for carbon sequestration, because it is extremely likely to burn.
In Australia nine trees out of ten bear the marks of fire; they have
adapted to thrive on particular frequencies and intensities of fire.
More than 355,000 square miles of eucalyptus forest—three-quarters
of the native forest area of Australia—stretch all around the coasts of
the continent.[25] *Eucalyptus regnans* is the largest species, a tree that can
grow to 330 feet tall and emits such a haze of flammable terpenoids
that its original niche, the Blue Mountains in New South Wales, is

---

* Wandering among the trees, I noticed a curious thing: the burns were all on one
side of the tree, normally the side facing up the slope. There are several reasons for
this, I found out. When a fire passes by a tree its height increases on the leeward
side of the tree, because two leeward vortices form. The flame height increases in
the vortices because the turbulent mixing of fuel and air is suppressed. The flow
of gaseous fuel in the vortices becomes greater than the rate of mixing with air,
and this forces the flame up; there is an increased height along which combustion
can occur. So fire scars are found on the leeward side of trees because the vortices
increase the residence time of the flame on the leeward side of the tree, compared
with the residence time on the trees' windward side.

wrapped in a lavender haze. *Eucalyptus regnans* trees show a unique coevolution with fire, carrying their seeds in capsules that can be deposited at any time of year. During a wildfire the capsules drop, and from these seeds a forest 200 feet tall can regrow in just twenty years. These trees will continue to stay dominant by frequently going up in smoke, and then growing again so fast that no other tree has a chance. Currently, 80 percent of Australian forest is eucalyptus dominated.

The story of its rise is that of humans and fire working in tandem. Eucalyptus had always been a pyrogenic tree, but when the climate grew warmer in the Holocene aboriginal people used fire to hold back encroaching rainforest, and as a result eucalyptus was favored by them and spread.[26] Charcoal and pollen records show a profound footprint of management on pre-European Australia, including the spread of eucalyptus along the Songlines, aboriginal routes described in epic songs.* One songline in particular, the Story of the Seven Sisters, starts in the Blue Mountains and goes down to the shore in the Dampier Archipelago, and the route now winds around the eucalyptus stands that spread through the highlands.[27] Aboriginal peoples cleared trees with fire, and eucalyptus replaced the slower-growing, fire-hating trees like oak.

In 1770, seeds from the eucalyptus were collected by Joseph Banks on the Cook expedition and brought back to the Royal Botanic Gardens in Kew, whence they were planted in the 1850s in California. They were wildly popular and in the 1960s spread throughout the tropics. Nowadays they are widely planted across the world, from Sri Lanka to Cornwall, and are the most common fiberwood source anywhere, as well as the most planted forestry tree. Not only do they grow fast, burn hard, and poison anything that tries to eat them,† but they will also lower the water table. The most beautiful eucalyptus stands I have ever seen are around the Georgian city of Batumi, where in the nineteenth century they were planted to make the soil drier. This water table lowering is an adaptation designed to kill trees

---

* In 1998 the Australian government employee Richard Bowman meticulously analyzed the charcoal record along the songlines.
† Only the koala can manage to digest their lethal cocktail of oils and soaps.

near them with drought while they grow fast, a little-known fact. In near-desert countries eucalyptus is available in tree nurseries as a quick-growing tree for greenery and shade; all too often the result ten years later is a scorched roadside and the death of other trees.

Humans assign benevolence to trees when it's often completely un-justified. I've talked to people enthusiastic about their tree-planting efforts in Cornwall, Kyrgyzstan, Iraq, Angola, America, and Tur-key, all of whom have decided that eucalyptus is the quick answer to reversing deforestation, apparently oblivious to the fact that it might backfire—and not in a controlled-burn sense. In *The Songlines*, travel writer and aesthete Bruce Chatwin describes waking up to find "the sun like a white blister, and a smell of burning." He recounts a watchful night in which "the rim of the horizon was on fire. The fire had been moving at 50 mph . . . The tops of the eucalyptus had been breaking off into fireballs and flying in the gale-force winds."* He was in Australia, but in Portugal, Sri Lanka, and Brazil fires like these have already started to happen, and they are likely only to get worse.[28]

How much does global warming encourage fire? The short answer is hugely—but the effect is not always easy to calibrate, partly because the record we have of earlier fires is inevitably patchy, and partly because changes in land use have a huge impact on the probability of somewhere burning. But it's not a situation without precedent. In the eleventh century, a time of warmth before the Little Ice Age, forest fires were common in the UK. "In this year," recorded the *Anglo-Saxon Chronicle* in 1032, "appeared wildfire the like of which no-one remembered before, which did great damage in many places."[29] In 1048 it once more records that: "The wildfire did much damage in Derbyshire as elsewhere," and then again in 1077, "This

---

* In Florida, speeds of 100 feet every second were recorded, and under the correct conditions of wind, fuel, heat, and drought, faster still may technically be possible. https://floridadep.gov/sites/default/files/FNAI%20Descriptions.pdf

was the dry summer and wildfire came in many shires, and burned many farmsteads and towns." Place names still record this: Brentwood, Burntwood,★ Burntheath, Burnt Hill, Brantridge, and Brant Fell in Westmorland.

Whether global warming will fundamentally alter the ability of trees to fire-regulate themselves is a pressing question because it has a bearing on how forests should now be managed. One of the last areas of old-growth forest in the US is the Yaak old-growth forest,[30] a soaring community of western red cedar, giant hemlock, shaggy sharp-needled spruce, western larch, lodgepole, and ponderosa pines reaching up to the sky and connecting to the 2.2 million–acre Kootenai forest. Three hundred thousand acres of this are now being logged, with the stated objective to "reduce catastrophic wildfire risk and improve forest health." For years various agencies have promoted logging to prevent fire. A thinned forest, they argue, means there is less fuel to burn. Others have fought back, arguing the absurdity of cutting down old trees and the likelihood that this will only enable younger, more pyrogenic trees to grow up in their place. Most eloquent is Rick Bass, who lives in the Yaak Valley and advocates for its protection as a landscape that has experienced minimal human interference. "This larch is not only meant to survive fire," he says, "it's meant to prosper from it. These attributes, the species diversity here, the structural diversity of the forest—they need to be studied, not clear-cut. But the forest service says that by clear-cutting a nearly 1,000-year-old forest, they'll teach it to be resilient?"

Excellent point. And yet across the globe there is evidence of trees' natural defenses and adaptations being tipped beyond mock fire to a point of serious, almost annihilatory fire that may existentially endanger trees as well as threaten humans. John Vaillant's marvelous and terrifying book *Fire Weather* recounts the story of the massive wildfire in 2016 that burned the city of Fort MacMurray in northern Canada, the heart of Canada's bitumen industry, in the forests of Alberta. These boreal forests are surprisingly modern. They started to develop about 10,000 years ago as the world came out of the Ice

★ Where the forest chronicle records a fire in 1296.

Age and the Wisconsin Ice Sheet retreated and assumed their current form only about 5,000 years ago, with birch and fir mixing into the black spruce, balsam poplar, aspen, and northern pine. They demonstrate that trees are not adapted to fire per se but to a certain set of circumstances, which they will themselves have partially shaped. Climate change, however, is causing fire-adapted species not to burn, but actually explode. Pyrotornadogenesis was not a concept until the Australian bush fires of 2003, when an extraordinary convergence of terrain and trees led to a fire that ripped sideways and formed a tornado a third of a mile wide, toppling a plantation of pines like matchsticks and breaking them off 10 feet above the ground. Not the kind of fire any tree would want. In the boreal forests great tracts have been burned, and fires can now be too hot for seeds to germinate, underground resprouting mechanisms can be lost, and spruce can explode . . .

A fire reverses the meticulous knotting together of the carbon chains that were themselves knotted together by the sun. The Carboniferous rainforest collapse 305 million years ago fragmented the tropical coal forests of Europe and America at a continental scale, so it's ironic that when scale-trees came out of the ground after their long burial in the form of coal they burned hot and through industrialization have been responsible for the clearing and burning of more trees than anything else and perhaps even the inauguration of a new fire age.

In 2020, camping in Iraq, my Kurdish friend and I decided to visit the Yazidi shrine of Lalish. Leaving Erbil late, we had taken a diversion to look for a stand of Syrian ash, *Fraxinus euphratica*, that had been recorded down the river in the 1980s but since seemed to have vanished from the landscape, so it was late afternoon by the time we drove up the narrow valley that protects the shrine. As we entered the valley, with its stream at the bottom and oak trees growing up the sides, we smelled smoke. Rounding the corner we saw a gas flare at the bottom of the valley, an eternal flame burning off the natural gas that comes up with oil, making carcinogenic compounds that give children cancer and wasting disease. Worse still, the hillsides were black with burnt trees, and the culprits were clear. Eucalyptus trees had been planted around the oil extraction station and along the stream and hillside, and they had sparked a fire that had spread into the oak trees and along the willows. The valley continued beyond the flare, turning left into a steep ravine, and after talking to the guards at the checkpoint we were allowed up to the main site. Lalish is large, with many temples, having expanded since the genocide in Sinjar as a safe haven for those who were forced to leave their homes. Yazidis were believed by nineteenth-century Christian travelers to be devil worshippers because they venerate a fallen angel, Melek Ta'us, and because they prize snakes, which they believe symbolize new life— their ability to shed their skins a sign of renewal. They also prize trees, an idea ISIS picked up on and, in an act of ecocide that compounded their genocide, cut down most of the trees in Sinjar.

In Lalish was a courtyard full of ancient beautiful trees, mulberries and pistachios, and people chatting. Beneath the ridged, conical spires was the main temple, with the tomb of the great Yazidi saint Sheikh Adi Ibn Musafir, who is considered an avatar of Melek Ta'us. The door was carved with suns and leaves and trees and a long black snake, and the custom is that you do not step on the high thresholds, so we went quickly in, past the holy spring and then downstairs to the vaults underground, black with soot and the oil of handprints. Large urns of oil stood in the shadows, and ancient pithoi were piled up further back. In the shadows was a holy fire of pine; there was a smell of incense, pith of a tree that had had fire and the adaptation to fire running in its veins, revealing these adaptations to humans only through the power of myth.

CHAPTER 4

Trees shaping air

How trees trap carbon, and the problems that this has caused

In an instant, like an insect caught by a spider, it is separated from its oxygen combined with hydrogen and (one thinks) phosphorus, and finally inserted in a chain, whether long or short does not matter, but it is the chain of life.

Primo Levi, *The Periodic Table*

When a huge dragonfly fossil with a wingspan of nearly 3 feet was found during open-cast coal mining at Commentry in France, it offered compelling evidence of the power of extinction.[1] It became an iconic image of the gigantism of insect life, a magnificent dragonfly gliding over a tree-fringed lake. Also trapped in the rock was an aviary of other gigantic insects, offering "such a striking series of strange forms as cannot fail to awaken the attention of the least incurious."* The dragonfly, along with mayflies the size of canaries, millipedes 8 feet long, and 9-inch cockroaches, came to characterize the Carboniferous period, a wet and warm period between the end of the Devonian, 358.9 million years ago, and the beginning of the Permian period, 298.9 million years ago. How could these vast insects have existed? Why were they no longer around? Debate soon raged back and forth about how the dragonfly (now known as *Meganeura*) would have been able to breathe, let alone fly.

The scientific community has now settled on a consensus.

---

* Paleontologist Charles Brongniart published his findings with this prefatory note written by Samuel H. Scudder of the *American Journal of Science*.

Meganeura is evidence of a time when oxygen was 35 percent of the atmosphere. This would have allowed the gas to diffuse into the large body of the insect and power its muscles in a way that simply wouldn't be possible today, when the concentration of oxygen sits at just 21 percent. And the reason the world's atmospheric oxygen levels were so high? Trees.

Trees take in carbon dioxide through stomata in their leaves, tie down the carbon into sugar, and give out oxygen as a by-product. They shape other elements of the air as well. They release volatile organic compounds, small chains and rings of carbon that not only seed clouds and shape water flow but also smell sweet to animals and allow trees to communicate with one another. They clean the air by trapping particles of carbon and sulfur from car exhausts and smoke. Exciting new research shows they also trap methane and alter the levels of nitrogen in the air through nitrogen fixation. Some of the changes trees make to the air are sporadic, varying over time and space and species. Nothing is as consistent as the way they lock away carbon dioxide and release oxygen, because the biomass of wood itself is twisted out of the air. When companies talk of carbon sequestration by trees it is comforting to know that what you see is basically what you get. In every square foot of tree you are seeing over a half pound of carbon dioxide: a world of 3D carbon in forests and woods. Millions of trees have been using sunlight over millions of years to solidify the air, and the result has been a rollercoaster of oxygen levels throughout deep time, affecting everything from the rate at which iron rusts to the size animals can grow. From a very low start point oxygen concentrations have occasionally become very high, sometimes to the point that trees themselves have had to adapt to survive by altering the very structure of their proteins and leaves.

It's an eerie reflection of the adaptations humans will have to make to high levels of carbon dioxide, and an inescapable irony that the proof of very high levels of oxygen produced by widespread forests 300 million years ago was found as a result of the extraction of coal. The elevated levels of atmospheric CO_2 we see today—almost double those before the industrial revolution*—come directly from the carbon of millions of years of tree photosynthesis, laid down 300 million years ago, locked away in rocks for millions of years, and then mined and burned in the last 200 years.[2] At the time coal was being laid down, humans, or their ancestors, were simple tetrapods like rather ugly newts, and the two great supercontinents of Gondwana and Laurasia were coming together to form the massive landmass of Pangaea, forcing up the Appalachian Mountains in America and the Hercynian Mountains in Europe. Pangaea formed from the north pole to the south and created around the equator a vast and perpetually warm and wet region perfect for the quick growth of trees and their collapse into swamps, where they formed coal. Since 1822 we have called this age the Carboniferous, from *carbo*—coal, and *fero*—carrying: the "coal-bearing" age.

Meganeura was introduced to science along with several other gigantic insects by Charles Brongniart, a paleontologist from a long line of fossil hunters and chemists. Between 1877 and 1894 he excavated a quarry at Commentry in central France that had been a narrow freshwater lake about 5.5 miles long and 2 miles wide, and in 1894 he published his findings in the *Bulletin of the Society of Industrial Minerals*. They caused a sensation. Soon the same huge insects were being found in other mines, where they had lurked unnoticed in tunnels deep underground.[†] Before long, in almost every place where coal was mined, from China to France, fossils were unearthed revealing evidence of giant insects and a hot tropical world. More and more turned up and continued to reveal different habitats and different modes of being. In 1979, at Bolsover in Derbyshire, miners found the Beast of Bolsover,[‡] another huge dragonfly of a different sort, the Protodonata, less a dreamy glider o'er the water than a fearsome dive-bomber plummeting between the trunks of a dense forest.

All the scientific explorations of the Brongniart clan and their dissemination across the Atlantic were funded and made possible by the

---

* Around 288 ppm before the industrial revolution to 424.55 ppm today.
† Rather than opencast mines.
‡ Not to be confused with the great politician Dennis Skinner.

innovations in mining that caused and were caused by the industrial revolution, which means that pretty much all our understanding of how trees have shaped the air for millions of years is probably also fueled by this dirty business. At this point in the nineteenth century the imprints of extinct plants and animals found in Europe were being linked up with living plants from countries in the tropics. In 1824 Alexandre Brongniart's 1821 article on "Fossil Vegetables traversing the beds of the Coal Measures" had been published in English in the *Annales of Mines*.[3] In it, Alexandre (Charles Brongniart's grandfather) notes that he was brought in by the heads of the mines in that area to look at the fossils. "These remains of ancient worlds," he writes, ". . . seem to have been so well preserved solely in order to furnish us with the only documents we could ever obtain on the natural history of these different periods."

He goes on to describe what we would now call lycopsids, giant clubmosses, upright in their original positions, and many tree ferns underneath them. Alexandre Brongniart recognized that the coal beds were "a true fossil forest," and that "all the vegetables of a tropical aspect must no longer then be sought to be found beneath the torrid zone, nor brought into our latitudes by means of great currents or grand debacles." He was echoing Leonardo da Vinci's assertion hundreds of years earlier that an intact ecosystem could not have been washed into the rock by a biblical flood, and concluding that the coal ecosystems were as they had been in their "native" rock. It was his contribution to two of the big debates of the time: whether the climate of France had once been tropical and the exact chronology of the rocks found in France and elsewhere.

Alexandre Brongniart was writing the year *before* the term *Carboniferous* was coined. The picture he built up of swamp forests and coal formation, a time when warm, damp conditions led to the disappearance of *Archaeopteris* and the formation of truly dense tropical forests across the globe, including in what would become Europe, came to define the period. The idea of fossils and geological periods was not new. In Italy, Leonardo da Vinci had observed fossils in the floods of the Arno River and had conjectured that, since the shellfish fossilized in the rocks showed no sign of disturbance, rocks that had once been at the coast must have moved lock, stock, and barrel inland. A hundred years later, Britain's Robert Hooke had recorded ammonites from the cliffs of the Isle of Wight, which he called "snake-stones" or "snail-stones," conjecturing that they were some extinct relation of the nautilus. The difference now was that Brongniart's vivid descriptions of what was preserved in the mines at Commentry were discussed everywhere—in America, in China—and converted millions of people to the belief that millions of years ago the planet's ecosystem was radically different, to the extent of the air itself.

Brongniart's meticulous work also contributed to the field of plate tectonics, which in the 1920s led to theories of continental drift and eventually to our understanding of the great slow movements of continents, of trees surfing across the world on land masses to fit themselves into ecological webs quite different from the ones evolution had designed them for. In the mid 1850s the idea of a supercontinent, Gondwana, had been put forward, but the evidence to support it was still being pieced together. Brongniart's discoveries slowly led to an understanding of just how old the world was and how much it had been shaped by trees. Time and time again, scientists discovered, climate change driven by living organisms had led to utter devastation. Living organisms could destroy themselves, although more species would arise, phoenix-like, from the ashes. When they defined the Carboniferous, ecologists drew a line at the end of the period marked by mountain formation and glaciation, which had dried out the landmasses and destroyed fertility, but even before then the huge tree clubmosses, the lycopsids, which had been the dominant trees in the big coal forests, had gone abruptly extinct.

Today, tiny clubmosses still grow around the world. In the UK they are rare and live on moorland, patches of bright green curling over the peat, tiny, soft kitten paws searching for new spaces to occupy. In the US they grow slightly larger. These are the minuscule descendants of the iconic clubmosses of the Carboniferous period: vast, scaly trees over 100 feet high, with stiff trunks that could reach a diameter of 3 feet. Their trunks were pith-centered and had two rings of xylem that were highly efficient at sucking up huge quantities of water from the floodplains they grew in and taking it up their photosynthesizing trunks to the spear-shaped leaves of the canopy. Once it was thought that, like lycopsids, they sped upward to find light twenty times faster than modern angiosperms and lived for only ten or fifteen years, but now this growth rate is considered physically impossible.[4] They must have lived for centuries, says modern science, laying down rings of stiff tissue like cork in a layer called periderm round the outside of their trunks for support, which they would have reinforced with woody lignin. In fossil form their trunks look like the traces of vast basilisks covered with scales and their cones like whiskery pineapples. Once upon a time they would have stretched in forests along the equator for green mile upon green mile, their feet in swamp and their roots with air spaces through them, their constantly falling branches building up the swamp bottoms again and again, occasional cyclones and fierce winds pushing them down to be replaced by a new generation.* Their fractal canopies, branching at different times in different species, would have cast angular, regularly dividing shadows on the water, quite unlike the dappled shade of a willow or an alder, and the *Meganeura* and other giant insects would have weaved in and out of their pillar-like trunks under a canopy of fuzzy green leaves. After reproducing they would have crashed down into the water of the swamp, and the roots of the next generation[5] would have grown into and through the soft and decaying pith of the old trunks until, buried and compressed, it carbonized and became the black coalfields on which we still rely today.

Vast clubmosses made up the majority of all biomass in the Carboniferous, particularly in coal-forming wetlands. Then, around 305 million years ago, these iconic trees started to die. The ground they sat in began to dry out, their thirsty structures crumbled from lack of water, and the ground they relied on became hard and cold. This is what is now called the Carboniferous rainforest collapse. After

---

* Sapling fossils are rarely found, just embryos and fully grown trees. To me this suggests that lignification came later, rather like in an elder tree.

just a few million years the 100-foot-tall green forests that stretched for thousands of miles had died of drought or failed to grow again from spores, and in their place only the squat tree ferns, 50 feet tall at most, were left, along with still smaller ferns and the cycads, making a feathery green covering over the earth. More drought, and these stunted forests in turn fragmented into refugia, small patches of green in a fast-growing desert. The air became drier and more arid. Glacial ice covered the south of the supercontinent Gondwana, and most of the huge continent dried out. In Cathaysia to the east, in what is now modern-day China, Korea, Japan, Laos, Thailand, Indonesia, and Malaysia, some Carboniferous forests persisted, and coal was produced in these areas until the end of the Permian, 252 million years ago, when they finally succumbed to the cold and a lack of water. Even now some of the tree species from the coal forest survive here, among them the forerunners of the ginkgos and cycads, which had diverged from one another during the boom in species at the beginning of the Carboniferous period. But the giant lycopsids were lost forever.

Most geologists agree that, although continental shift influenced the climate, trees themselves were a major reason for the demise of the lycopsid forests. The amount of oxygen produced by trees, and more importantly the amount of carbon dioxide consumed to create the tree biomass that was subsequently locked away as coal, had, they claim, changed the atmosphere to an extent that led to dramatic global cooling. But even with forests covering much of the earth, how did carbon dioxide levels plunge so low and oxygen levels climb so high? How did trees shape the air so comprehensively? How did they trap so much carbon away? And in a world where we are trying to lower carbon dioxide levels again, what does that mean for us?

Wood as we know it is everywhere now, creating amazing structures greater than cathedrals: the lofty pillared beech woods of Romania, the mazed chestnuts of Wisconsin, or the crawling, many-layered complexity of the oak woodlands of the Helford River in Cornwall.

But it had a strange and painful birth. To understand its relationship to air, and therefore how trees have shaped the air, we must follow a molecule of carbon through the mazes of millions of years of chance, as it was changed from air into solid in the form of sugar; from sugar into the Teflon-like carbon substance called sporopollenin, and sporopollenin into lignin, the stiff, inert carbon backbone that supports every tree.

Carbon dioxide is a flighty little molecule, two oxygens both surrounded by negative charge, forced to share a carbon. The result is a molecule shaped like a tortured insect, the two constantly vibrating oxygen wings pulling away from each other and staying at a painful 180 degrees with a carbon thorax in between holding them both in place. The green pigment chlorophyll★ traps energy from sunlight, and plants use a protein called Rubisco,† the most abundant protein on earth, to hold carbon dioxide and use energy from the sun to force its two oxygen atoms together so they combine and spring away into the atmosphere as oxygen gas. The carbon, meanwhile, painfully and awkwardly, is linked to another carbon to become part of a carbon chain: the backbone of sugars, starch, and the majority of life. Once the carbon-carbon bond has been formed six times, trees have effectively made the six-carbon chain of glucose. After it has happened twelve times they have made sucrose. Normally, respiration in animals, fungi, and bacteria would break down these sugars, releasing the energy trapped in these molecules and combining the carbon chains with oxygen to recycle the carbon as carbon dioxide back into the atmosphere. So why didn't this happen to the trees of the Carboniferous?

Five hundred million years ago or earlier, when the first green cells were washed up on the beach, they were destroyed by the sun. If you look at the seashore after a storm you will often see green lines along the sand, gradually turning black. In the same way that humans suffer from sunburn, the sun's energy, particularly ultraviolet (UV) light, causes havoc in this algae's transparent green cells no longer protected

★ And, of course, other pigments too.
† Ribulose bisphosphate.

by the sea. The UV light charges up high-energy electrons that attach to oxygen and form highly reactive chemicals called free radicals. It's a good name, because radicals can cause reactions between
substances that usually would never combine. In particular, they can
twist carbon chains around until, like the mythical serpent the ouro-
boros, they swallow their own tails. Everything changes, unpredict-
ably, drastically, when UV light from the sun sparks off free radicals,
and even genetic material designed as a blueprint that would sit safely
at the center of cells is split up and destroyed.

Repeatedly, over millions of years, cells of algae from the sea that
washed up on dry land were destroyed, until one lucky accident
sparked off a world-changing free radical chain reaction. This reaction, probably between the carbon chains of fats, produced an irregular mass of carbon rings and chains called sporopollenin, which
soaked up UV rays and free radicals like a sponge. The green cell
that was first capable of withstanding the sun on a dry beach was
an alga, of which today chlorella is the nearest living relative. With
a protective shell of sporopollenin, finally a photosynthesizing cell
survived on land, and 470 million years ago the photosynthetic cells
protected by this chemical, a powerful antioxidant, developed into
the first land plants.

Plants learning to survive and use UV light was a thunderbolt, a
Promethean vista that allowed a whole new chemistry to emerge,
root, and branch, in a whole new place: dry land. The result is the
shade of forests and the shelter of woods we now enjoy. Safe from
predators, who for the moment were left in the sea behind them,
these photosynthesizing cells started on a path that led to the amazing complexity of trees, from the Amazon rainforest to the baobabs
of Madagascar and our own habitats. Free radicals are better than
anything else at carbon-ring chemistry, and carbon rings allow an
amazing complexity of forms. At the molecular level it is like jumping into a ball pool and suddenly having tennis rackets to play with—
the world becomes a totally new and much more exciting place.
Soon, plants were using carbon dioxide to make carbon rings that
could handle tricky chemistry, the forerunners of what we now call
aromatic compounds: chemicals that smell sweet to our noses, the

smell of grapefruit and pineapple, cedar and lavender. These chemicals could fight herbivores, attract pollinators, materialize rain, and summon fire, some of the most fundamental chemical reactions of life.

Four hundred million years ago there were no herbivores and no pollinators, but for the earliest green cells on land reproduction was a major struggle, and the weather was violent: tornados, floods, and biting winds. Sex wasn't much on their minds, but under threat the early cells packaged themselves into hard round balls we now call spores. The rest of the time they replicated themselves by dividing whenever they could. Sporopollenin, the same complex mess of carbon chains that acted as body armor for algae against the sun, became the robust packaging that protected spores. A carapace almost indestructible, it allowed genetic material to move from one place to another and withstand drought, which gave cells the capacity to make sporopollenin spread. As those early cells became more and more successful they started to organize themselves to grow up toward the light, and finally, about 400 million years ago, the first woody plant appeared.

What enabled it to grow high was a stiffness, like a skeleton, inside it, which allowed water to flow up from the ground to its leaves. That skeleton was formed by lots of bits of sporopollenin joining together into a longer, stiffer, hydrophobic, and unreactive polymer—lignin. By the time the first trees, *Archaeopteris* and the smaller versions of lycopsids or clubmosses, reached 26 feet in height, lignin made up the very fiber of their being, stiffening each layer of old cells as the tree grew up and the trunk grew thicker. At the end of the Devonian *Archaeopteris* died out, but lycopsids continued to grow and develop, to move over every part of the hot, warm, wet earth they could find, and the result was lignin on an enormous scale—100 billion tons of carbon locked down and then compressed and hidden and carbonized into coal.★ Two hundred million years after lignin first appeared on earth, *Meganeura*, the dragonfly with a 3-foot wingspan, could fly, breathe, and feed, all in 35 percent molecular oxygen.

★ CO_2 concentrations plummeted tenfold from 2,000 ppm to 200 ppm.

By the Carboniferous period early herbivores, the synapsids,* of which human ancestor tetrapods were a part, did exist, but trees, in stiffening themselves so they could grow taller and encourage water to rise from the swamps at their feet to the leaves 100 feet up, had given themselves the equivalent of a crab's shell: a tough, inedible material we would recognize as the essence of wood. Lignin was tasteless, odorless, fibrous, and remarkably hard to digest.[6] A strong crosslinking of a hundred long carbon chains and the boring, pedestrian tangle of unreactive carbon bonds, handcuffed together with smooth carbon rings (as well as an occasional safely bonded oxygen atom), makes it almost impossible to break down.

For early bacteria and fungi, breaking down lignin would be like trying to split logs with your bare hands—they simply didn't have the tools. We can still see the resilience of lignin in everything made from wood today. It's there in Viking longboats and medieval church doors, modern joists and the beams of tree-houses; in old oak chests and garden sheds; in wooden boats buried deep in mud and the exquisite carvings of Grinling Gibbons. It's also in the bonds that (along with cellulose) bind the paper of this book together and stop it from gently crumbling in your hands.

Enzymes normally oozed out to digest plant material but were unable to get their teeth into anything. There was no juicy metal ion or nitrogen atom to offer a polarization of charge, and the general "fit" of enzymes was not suited to something so irregular.† And because lignin offered a big competitive advantage in the early forests, stiffening xylem and phloem and allowing trees to grow high and transport water up to their crowns, it quickly boomed, and there was a lag of many years before anything was capable of degrading it. Even today there is only one organism that can really break down lignin: the white rot that fungi use to digest wood. The key to breaking it down is the same as to building it up—free radical chemistry—but in the

---

* Synapsids are a diverse set of animals including all the proto-mammals.
† This is an ongoing debate. Some ability to break down lignin did exist, but the major white-rot enzymes that fungi use to break down wood appear not to have been there.

Carboniferous period, although white rot might just have evolved, it was not significant enough to make a dent in the piles of wood. Only fire could consume these, and until humans started acting as fire-feeders even fire was impotent. This allowed fallen and living tree trunks to be buried intact, so the carbon in them was locked away. Meanwhile, the oxygen pumped out by all those trees was not consumed at anything like the rate we consume it now. There were fires in the wet woods, but not the million fires of a world with humans in it. The rocks that had been exposed were weathered and had already oxidized, the animals soaring through the oxygen-rich environment were, with their shallow breath, unable to make a dent on the overall picture.

The carbon locked away by trees remained in the ground for millions of years, until about 3,000 years ago, in Fushan, from a mine operated by the Chinese Han dynasty, humans started burning coal as fuel. The industrial revolution around the world ratcheted this up on a massive scale, with the results we are familiar with today. Even so, the percentage of carbon dioxide in the atmosphere still stands at only 0.04 percent, and the oxygen level of the atmosphere is practically unchanged by humans. An estimated 1.16 trillion tons of coal is still locked away safely in the ground, mostly in the US, but also in Russia, Australia, and China. For trees to change the oxygen concentration of the atmosphere they had to master the chemical steps of billions of years of evolution and use them to create a whole new world. At the end of the Carboniferous earth's temperatures fell in some areas as much as 36°F, and with glaciation came intense monsoons, raging wildfires, a drop in sea levels, and a general acidification. These changes happened over millions of years, but many tree species were powerless to evolve away from their ultimate destruction.

Trees shape the air globally, changing concentrations of carbon dioxide and oxygen over long time periods. As they do, they also shape the air at a more micro level, in the airflow through forests and the invisible blanket around a tree. The earth is like a great, lazy spinning top that sets up airstreams as it spins, broadly west to east, apart from the countercurrent of the trade winds in a ribbon round the equator, but closer to the ground the prevailing wind is often shaped by trees and has been since the first trees grew up on coastal plains. In cities this phenomenon has been closely studied, because trees prevent wind tunnels forming and clean the air by trapping dust and exhaust particles.

Many trees, the poplar and willow, birch, hornbeam, and sycamore among them, use wind to disperse their seeds. This is called, beautifully, anemochory. Anemochory enables the very light fluff of poplar and birch seeds to be whisked huge distances on autumn storms to patches of wasteland or disturbed floodplain where they can grow up without competition. For this reason they are called pioneer trees, and it has been a favorable tactic, particularly across the huge landmass of Siberia and Mongolia, where forests consist of larch, pine, and birch, but in the barest areas and along rivers it is willow, Siberian birch, and poplar that dominate and survive, miles from the next nearest tree.*

Trees like ash, elm, and sycamore that are slightly more tolerant of shade and competition have samaras, heavier seeds attached to wings that spiral out from the tree as they fall in circles, like the spiral minaret of the great Abbasid mosque of Samarra in Iraq. They are also known as keys or spinning jennies, because as they twirl away from the adult trees they spin like helicopter rotors. This makes the majority of the seeds travel only a little farther downwind of the adult trees than they would if they had no wings at all, and so their progeny will grow up in the shelter provided by their own parent trees and become as tall as them. Left to their own devices across a plain, this means the gradual expansion of these trees in a westerly direction, a sort of cavalry charge of trees in the shape of a shield, growing up in shelter and then reinforcing it. When the wind is not too strong

---

* Two smaller trees—*Paliurus spina-christi* and *Combretum zeyheri*—shape the air across savanna environments so that the seeds can roll. *Paliurus spina-christi* was supposed to be the tree from which the crown of thorns was made for Jesus to wear.

a few seeds will also travel a little upwind of a tree, so the shelter of the trees downwind doesn't just hold but increases, and the trees do not lose ground.

This evolutionary strategy reaches its apogee in the tulip tree, a tree genus, *Liriodendron*, with only two species, one of which has green, cup-shaped flowers and is native to China and the other, with larger yellow flowers, native to North America. Both have large, distinct leaves with four lobes, something between a heart and a heavy-hipped fertility goddess with pointed shoulders. While some trees are semi-evergreen in particularly sheltered nooks, most will lose their leaves in the winter. In temperate forests in North America wind blow causes half of all tree deaths, and the major factor is the "sail effect"—the resistance of leaves against the wind. In strong winds the shape of tulip tree leaves allows them to curl up into a tube, thereby lessening the wind's force. The seeds, on the other hand, are like ash keys, designed to fly a good distance.

One of the most impressive collections of tulip trees outside their native habitat in the Appalachian Mountains is in the New York Botanical Gardens, a haven of old-growth temperate forest and extraordinary plants just next to the Bronx. Its iconic avenue of huge trees reaches up to a high canopy. Planted in 1903, the trees are now toppling over and causing problems. In fact, they were never meant to be outside a sheltered grove of their own kind and rely on seed distribution patterns to make this happen. The North American species can grow over 165 feet tall (the star specimen is 190 feet, a few feet taller than the Leaning Tower of Pisa), so it's important for the seeds to travel a good distance if they are to find an area clear of their parents' canopy. In 1890, John Loudon, a Scottish botanist, noted that seeds from higher up the tree were more likely to germinate, and the reason seems to be that these are likely to land further away from the tree, so less likely to shade out the parent.

Tulip trees are becoming more and more popular as street trees in Europe and America, and while the American tree with its large yellow flowers is the more common, the Chinese green-flowered tree is popular too. Like any street tree, they are valuable not just for the shelter they produce, which reduces the wind-tunnel effect of streets, but also because their leaves can attract and soak up particulate matter from polluting exhaust fumes and smoke. Pinning down why trees remove particles from the air so actively is a field of very intense research. Trees with hairy leaves in particular, like the downy birch, or even the plane tree, can clean a street of its particulates much more effectively than anything else.

The irony is that, having removed particles from the air, trees then add their own. Almost 40 percent of people in Japan now suffer from hay fever,[7] to the extent that they cannot go outside or sleep without medicine.[8] One reason suggested for the sudden appearance of trees in the Devonian era—characterized by a sudden growth spurt—was that being taller allowed them to release their spores from high up, thereby encouraging the spores to spread further and give them an evolutionary advantage. Trees use the air to move their genetic information around, designing pollen to be as light and small as possible—just genetic information under a tough sporopollenin coat—so it can fly on airflows for thousands of miles. Under a microscope the tiny particles of sporopollenin are shaped into ridges and patterns that are species specific, allowing trees to identify their own pollen if it lands on a female flower. Pollen tends to be negatively charged, so it will attach to the female flower. Unfortunately, this combination of pattern and charge is exactly what the human immune system recognizes too, and in launching a fierce attack on what it thinks is a pathogen, it can reduce sufferers to streaming wrecks.

In many countries around the world pollen causes problems, but the issue is particularly acute in Japan because after the war a huge tree-planting program was pursued, which included the establishment of huge numbers of cedar and cypress, including sugi, *Cryptomeria japonica*. Because of global warming these trees are now maturing earlier and producing more pollen. Compounding the problem is a historic bias toward planting male rather than female trees as, expending little energy on reproduction, they tend to grow slightly faster and look more robust. Even trees that are normally hermaphrodite tend to produce more pollen at the beginning of their lives, maturing only at the end. Luckily, in sugi about two in every

8,700 sugi trees have a mutation,* which means the pollen coat of the tree does not form properly and the tree does not release pollen. This is now coming to the hay-fever sufferers' aid.

Most conifers will saturate the air with pollen. Sporopollenin, the same compound that protected the first cells from UV light and agglomerated to form the lignin of wood, is shaped in massive quantities into capsules with mainly asymmetrical spikes and sizes, which are then filled with DNA. For a few days in the year small structures at the end of pine branches will produce huge amounts of powdery yellow pollen. Under a scanning electron microscope pine pollen looks like a pair of headphones or a crab with its claws tucked up. A larger, rounded cap has two smaller roundels underneath it, effectively balloons filled with air, and the pine germinates cells that expand from between them. When the tree releases pollen these roundels act like wings, allowing the pollen to ride the wind or an airflow over a massive distance. This has probably been a major contributing factor to the success of the pine family, which produces so much pollen it can be seen from space.[9] In China and Korea pine pollen is sold in bulk as a topping for rice cakes, and to make a delicious form of fudge. Slicks of pollen on the Baltic show up on satellite images; it is a major marine food source, and pollen can be found on the ice of the Canadian Arctic, 1,865 miles away from its source.

Self-pollination is a last resort for trees, a necessary evil when cross-pollination and the consequent genetic variation have failed. Compared with angiosperms, conifers grow widely spaced, ensuring a throughflow of air around their branches. This is reflected in their architecture: the araucarias of New Caledonia branch and whorl in such a way that air will first move past the bottom of one tree in a stand and then be whirled up to the top in a sort of vortex. The male cones are at the bottom of the tree, the female at the top, which is thought to reduce the chance of the tree pollinating itself. All the research into street trees in cities reducing the wind-tunnel effect agrees that they are essential, better than any other method of wind

---

* In a gene that scientists have called *MS1*.

reduction[10]—but there are very few studies of the shelter-belt effect of all the trillions of trees on the planet.[11]

Trees shape the air globally, changing concentrations of carbon dioxide and oxygen over long time periods. They also shape the air at smaller scales: in the airflow through forests and the invisible blanket around a tree. Finally, and perhaps most crucially, they shape the air within themselves in the heart of their own most fundamental compartments—cells. As well as theorizing about the early fossils, Robert Hooke, the great technician of the Royal Society, whose book *Micrographia* showed the natural world in unimagined detail, turned his primitive microscope onto a thin slice of cork and, as a result, in 1663 he described cells for the first time. He chose the name because (as he wrote in the *Micrographia*):

> the interstitial or walls (as I may so call them) or partitions of those pores were neer as thin in proportion to their pores, as those films of wax in a Honey-Comb (which enclose and constitute the sexangular cells) are to theirs.[12]

Cooperation, rather than the incarceration overtones of *cell,* was clearly on his mind; had he looked at a living, green cell from a cork-oak leaf he would have seen that there were prisoners locked into those cells—that all the green of the leaf was held in a mass of tiny floating capsules that replicate individually and migrate toward light. These were chloroplasts, and they were once free-living cyanobacteria, tiny cells a bit like blue-green algae.

Trees require chloroplasts, these tiny subcellular structures, to split off the oxygen from carbon dioxide and fix down the carbon—the tree-wrought atmosphere may never have become over 30 percent oxygen were it not that trees shape the atmosphere by proxy. Just like cows or humans, which enlist bacteria to perform the chemistry they cannot handle, trees and plants use bacteria to manage the chemistry they can't grasp, including manipulation of the smallest gases such as nitrogen, oxygen, and carbon dioxide. Like people living in societies where rubbish is whisked away and can therefore be ignored, chloroplasts were insulated by trees from the effects of the oxygen they were producing and could get on with fixing carbon and driving up oxygen levels. Rather than reacting to the air concentration as it was at that moment in the atmosphere, chloroplasts were, and still are, photosynthesizing carbon dioxide at the rate that trees contrive. The result is that trees can leverage up concentrations of gas beyond the point that would be in their own best interests—and in the Carboniferous as high as 35 percent.

The irony is that chloroplasts entered larger cells to *escape* oxygen. Like the protagonists of Doris Lessing's book *The Making of the Representative for Planet 8*, they are the sublimated survivors of an apocalyptic snowball planet. Billions of years ago, as free-living cyanobacteria, their ancestors produced so much oxygen that they almost destroyed life on earth. These tiny membrane-bound cells were green and photosynthesized; they had pigments in them that trees still use, like chlorophyll and carotenoids, and they had membranes proteins could embed into and use as a sort of battery.[13] They were so successful at photosynthesizing on the high–carbon dioxide, low-oxygen earth that they boomed for millions of years, oxidizing the planet and finally forming so much oxygen that they lowered global temperatures enough for the earth to be covered from pole to pole in ice sometimes miles thick.

The early cells used metals like iron or magnesium to get a grip on gases; these had a looser relationship with some of their electrons than carbon or nitrogen and could gain or lose electrons easily.* The first organisms had fought to become ordered, but to control gases a tiny element of disorder was necessary; just a soupçon, like an untidy kitchen in a very tidy house. What the first photosynthesizing bacteria needed was an awkward combination, something that could hold onto gases strongly enough that they would react, but not so strongly that it would never let go—a sort of reversible grip, a very early equivalent of an eyeshadow[†] or an electromagnetic lifting magnet.

---

* Take a copper wire, for example, or an iron bar; both will be able to conduct electricity because electrons can be pulled from one end to the other, but the same thing is not true of something made predominantly of carbon, like a tree or a person.

[†] In waterproof eyeshadow, the magic binding substance is normally zinc stearate, a sort of soap that clings to the fat on your eyelids until it's wiped away by a more oily oil.

The answer, 3,500 million years ago, was the porphyrin ring, a carbon ring with four pincer nitrogens, like a UFO grabber, that hold and stabilize the metal ion. The most famous modern example of this structure is hemoglobin, which binds oxygen in our lungs and releases it to our most distant cells. In this essential metalloprotein iron is held in the nitrogen pincers and stabilized by protein, but in the millions of years of evolution there have been hundreds more examples. Early photosynthesis was basic: a porphyrin ring could absorb light and use the energy to join carbon to carbon, and all the trimmings of photosynthesis as we see it now—the paths of gases in and out of the cell, even the magnesium ion at the center of chlorophyll—are the tweaks made by the giant hand of evolution over millions of years. Still, that original porphyrin ring structure worked so well that we use almost identical structures constantly today.

Under the ice, some cyanobacteria hung on between ice and rock, photosynthesizing slowly or ekeing out any tiny piece of energy they could from the earth. When cyanobacteria are stressed, their response is to take up DNA from the surrounding area. Useful genes quickly spread and, as the earth heated up again, bacteria of all kinds boomed, but oxygen and free radicals were lethal for the cyanobacteria, much as they were for the earliest land cells. The result was that 1,200 million years ago in the sea, a cyanobacterium nestled up to another single-celled organism, an archaea, which could use oxygen, and eventually their association became so close that the cyanobacterium was progressively engulfed. In return it was a source of hydrogen ions for the archaea, finally becoming the chloroplasts we see today.* Another similar merging, this time with a purple non-sulfur bacterium wishing to escape oxygen, produced a cell with both chloroplasts and mitochondria, a tiny capsule able to digest sugar and produce energy for the cell, and the photosynthesizing eukaryotic cell was born: the cell found in every leaf of every tree in the world.

---

* And by extension the chromoplasts, etioplasts, leucoplasts, and perhaps even tannosomes.

What does this mean for trees and the air? Well, both photosynthesis (consuming carbon dioxide and producing oxygen) and respiration (consuming oxygen and producing carbon dioxide) were now locked in the cage of a more sophisticated environment, an environment, that could feed them, seeking out and providing carbon dioxide and sunlight, but that would then harvest the results and never again let carbon, once fixed, go. The engulfed bacteria were now manipulated by something bigger than themselves. One can come away not liking eukaryotic—a cell that contains membrane-bound organelles—very much: they immobilize bacteria and enslave them, removing some of their important genes so they can't replicate without the permission of the host cell. In this light, a tree, rather than the liberatory mechanism we've been thinking of, becomes a vast tower block of enslaved photosynthetic symbionts—chloroplasts stripped of most of their genes, mitochondria stripped of their flexibility, a dystopian high-rise as in the eponymous J. G. Ballard novel, with its inhabitants working, working, working for a hive. However, a tool, particularly a living tool, cannot be used without using in its turn.

Like an ancient and deadly blood feud, cyanobacteria competing for useful gases have carried their billion-year-old conflicts into the heart of trees. The person who worked out that the little shapes in tree leaves that had once been free-living bacteria were responsible for the way air was shaped by trees was Lynn Margulis, a precocious biochemical genius by any standards. She had finished her bachelor's degree by eighteen and by the age of twenty-four had a doctorate from the University of Berkeley and was on track for the publication of her greatest theory. Her paper on an endosymbiotic—literally "an engulfed and mutually reliant"—origin for chloroplasts, mitochondria, and the motile flagella of some cells was rejected by fifteen journals before it was finally accepted and published by the *Journal of Theoretical Biology*.[14]

Her endosymbiotic theory stated that eukaryotic cells, the great lumbering tanks of complexity that are the basis of yeast and humans, antelopes and oak trees, are the result of bacteria being enveloped and enslaved by other cells, and therefore we can understand eukaryotes better if we take bacteria and their constant, fast, flexible use of gases in metabolism into account. Her overarching philosophy was that variation in living organisms occurred via symbiosis and cooperation rather than competition.*

She was fiercely criticized, but before too long the rapidly advancing fields of molecular and cellular biology had triumphantly confirmed her thesis. To the amazement of many, a ring of DNA in the chloroplast replicated as the chloroplast replicated. This circular chromosome, typical of bacteria, didn't have all the genes chloroplasts needed to live and function, but it had the one that really mattered: the gene for a bacterial version of the photosynthetic reaction center, the key to light capture. Clinching the evidence were the protein-making ribosomes of the chloroplast—bacterial copies, smaller, and with a slightly different structure—less good at editing the chains they made. Margulis had triumphantly proved that the cell was not just a cell, it was a metamorphized bacterial metropolis.

This makes a large tree a miracle of balance between opposing forces. Each leaf will have roughly 30 million cells. In each leaf cell will be about 40 chloroplasts and probably 300 mitochondria, and within the same cell these internalized, adapted bacteria are doing completely opposite things. One is building things up, and one is breaking things down. One will consume oxygen and give off carbon dioxide, while the other consumes carbon dioxide and gives off oxygen—and these demands must be balanced in an organism with no central control system we know of.

Trees live in a blind world that is exquisitely sensitive to the air that comes into and through the stomata. Every year 40 percent of the

---

* Other papers that Margulis has gone on to write have been less warmly received but may yet be proved correct. One of these was "On Syphilis and Nietzsche's Madness: Spirochetes awake!," in which Margulis argues that just as humans are covered by bacteria and influenced by the bacteria in their gut, there are many examples that we ignore in which symbioses are happening right under our noses. There are the obvious ones—the lichens, for example—and then there are the less obvious ones, which are still tenuous possibilities rather than anything else.

carbon dioxide in the atmosphere passes through stomata.[15] We have already seen examples of stomata controlling responses to smoke and water loss, but it ratchets up a level when it comes to adjusting to carbon dioxide levels inside the cell. In 1987, experiments comparing the stomatal levels of trees in herbarium collections and current trees in the south of England showed that the leaves of trees in southern England today have 40 percent fewer stomata than 150 years ago,* an extraordinary tree response to human-driven increases in the carbon dioxide levels in the atmosphere. Trees don't want more carbon dioxide than they can handle inside their leaves, so they use a mechanism that is beautiful in its simplicity. High carbon dioxide levels in the leaf switch on a gene called *HIC* or *High Carbon Dioxide,* which encourages deposition of fatty acids in the stomatal pore cell. This inhibits neighboring cells from developing into stomata, limiting gas exchange that is not essential.†

Years ago, my big solace in the face of climate change was that having more carbon dioxide in the atmosphere would be an advantage for trees, allowing them to grow faster and better and thus counteract the human-wrought increase in carbon dioxide. It was a simplified version of James Lovelock's Gaia hypothesis, that the earth acts to heal itself, and I think many people have or had similar consolatory half-beliefs. We have, after all, the evidence of the Carboniferous: however dramatic the atmospheric change, something can come out of it alive.

Throughout the 1960s, NASA, terrified by the threat of acid rain, ran missions to visualize the gases in the air: carbon dioxide, oxygen, nitrogen, nitrogen oxide, sulfur dioxide, and ozone.[16] One of those to participate in the study was Lynn Margulis. She proposed that the bacteria on earth and in the sea had an extraordinarily important

---

* The actual number is 67 percent over 200 years, but we have followed the lead of David Beerling in *The Emerald Planet* for clarity.
† Without this gene, stomatal pore-making spirals out of control.

effect on the atmosphere, a radical view in a system that had been designed to measure the human outputs of carbon dioxide and the effect that sulfur dioxide had in making acid rain. Acid rain had burned forests and denuded hillsides, but Margulis knew, in more detail than her colleagues, who were mainly geophysicists, that for some bacteria, a drink of acid rain is like a draft of citron pressé, an energy-giving, life-enhancing thing. When acid rain fell, bacteria that consumed sulfuric acid to get energy would boom and very soon would use up all the acid in the system. Nature would correct itself.

Margulis knew that the metabolism of bacteria was so flexible that almost anything humans could throw at the environment would have a bacterial target, and that these bacteria would rapidly multiply and counteract the change, bringing the system back to equilibrium, and she felt that as a result we can rely on trees—as we would rely on bacteria—to maintain some form of dynamic equilibrium in the atmosphere.

Margulis and Lovelock coevolved the idea of the Gaia hypothesis along these lines in the 1970s, explaining that the earth is a self-regulating complex system, in which organisms will expand into niches or streams of abundant food, consuming the energy in it and returning the environment to equilibrium as a result. James Lovelock was at pains to explain in later books that the rate at which humans were pushing climate change could never be counteracted by the natural world, but for Margulis, humans might perish, but trees and most of the natural world would always find a way to adapt out of or around any climate disaster that humans could make.

In 1997, a team of researchers analyzed satellite data to reveal a dramatic stimulation of forest growth in one of the warmest periods of the past 200 years, from 1981 to 1991. The spike was caused by a combination of heat and the carbon dioxide fertilization effect, the way in which higher concentrations of carbon dioxide allowed Rubisco to work more efficiently and incorporate more carbon into wood without increasing the need for nitrogen or phosphate.[17] In young forests this effect is marked and has been well attested in climates as different as those in Sweden and Brazil, but in a paper published in July 2024 even an old oak woodland in Staffordshire in the UK was found to be affected by higher carbon dioxide levels.[18] Growth of wood increased when carbon dioxide was increased, thereby locking down carbon more quickly than before and, contrary to popular fears, it was not cycled quickly back up into the atmosphere again, but stayed locked down for good.

This hopeful sign of trees responding to correct the heightened levels of carbon dioxide that humans have released by burning fossil fuels is compromised by rising heat levels around the world. Not only are forest fires on the rise because of drought and changes in weather patterns, but heat also more than counteracts the effect of heightened carbon dioxide levels. It causes photorespiration to roar into action, making it hard for trees to survive and reproduce, let alone lock down more carbon than before. Heightened carbon dioxide levels are too small to stop raging wildfires brought on by heat and drought and changing weather patterns, and there is no way the huge amount of deforestation still happening around the world could be countered by a simple rise in wood production. Nonetheless, the ability of trees to shape air shouldn't be underestimated. They may not do it on a human timescale, but given space and time trees will steadily soak up so much carbon dioxide that they will, as they did in the Carboniferous, fundamentally shape the air.

A recent big surprise has been that trees have an even greater role in preventing global warming than we first believed. Trees were thought to produce methane—but it turns out they actually *consume* it. Gases and their fast, invisible movements are hard to unpick, particularly at a global level. When NASA first developed its monitoring systems for CO_2 it ignored methane, CH_4, a gas we now know to be one of the most significant drivers of climate change, accounting for 30 percent of global warming. Methane is produced by bacteria in the absence of oxygen, in swamps and mires and dark wet places. Wetlands like the Amazon can produce huge amounts of it, as do coal mines and oil wells. Trees were thought to release small amounts of methane by acting like a chimney, taking it up from the soil and then releasing it slowly through their trunks, and this remains true for the first 3 feet of a trunk. Beyond 7 feet, however, it seems that bark bacteria consume so much methane, particularly in tropical ecosystems, that they end up taking methane out of the air once they've consumed all the part coming out of the trunk. Trees (or their associated bacteria) appear to be responsible for 26 million to 55 million tons of methane uptake globally.

A new project has started to monitor methane at two levels— broadly in the upper atmosphere and in detail over the swamps and bogs of the Amazon, the gas flares of Iraq, and the rubbish tips of Canada. What it has shown so far is that the carbon sink in global forests is steady, but the ability of forests to respond flexibly to changing atmospheres shouldn't be oversimplified. Trees will probably act to maintain dynamic equilibrium, but this relies on them having space to adapt, while the lessons of deep time tell us that gases are tricky little blighters, and you never know who to trust.

CHAPTER 5

Trees shaping fungi

*How trees use fungi for their own advantage,
and how they sometimes lose*

Let sea-discoverers to new worlds have gone,
Let maps to others, worlds on worlds have shown,
Let us possess one world, each hath one, and is one.

John Donne, "The Good-Morrow"

In the New York State Museum in Albany, where I went to see the first fossil-forest trees from Gilboa and Cairo, they have a prized fossil of a *Prototaxites*, a vast and trunkless leg of black stone, featureless and strange, lying in the corner. Reconstructions have shown *Prototaxites* branched like conifers, with mycelium networks under the ground. The name comes from one of those common early fossil misidentifications. The man who first studied these black monoliths,* the Canadian John William Dawson, described them as partially rotten conifers and thought that the microtubules he could see under the microscope were the fungi that had been decomposing them. He called them "early yew-trees," *Prototaxites*, and the name has stuck.

Although at almost 30 feet tall *Prototaxites* towered over any other plant known from this period (the nearest in height was probably the 2.5-inch-tall *Cooksonia*), it was already taking nutrients out of plants by putting mycelium into them. You can see the fossilized mycelium

---

* The first was found in 1843, and after that they appeared everywhere. Eukaryotic cells (just like plants), fungi had formed vast *Prototaxites* when trees were still struggling to grow up.

in fragments of the Rhynie chert, Devonian era fossils from Scotland
that show immaculately preserved plants close to liverworts, as well
as the fossils of *Prototaxites* itself. Later, more mycelium networks can
be seen in Devonian trees, in the first ectomycorrhizal interactions in
pine from 156 million years ago, and today most trees form some sort
of associative network with a fungus or two or even three or more.[1, 2]
Like most of our own relationships with other species, these are not
relationships of simple benevolence or mutual support. Although
some fungi will kill trees and some not harm them at all, most sit at
an awkward interface, a balancing point between support and harm,
where the forest network can turn on a coin and the fungal support
network quickly change into the digestion matrix. Trees use fungi
for their own advantage, but they sometimes lose.

Fungi break down and grab nutrients from other life-forms.
They would not be here without photosynthesizing organisms,
alive or dead, while trees, although they often benefit from fungal
networks, probably would be here without fungi. We know this for
two reasons. The first is that most trees will grow happily in a sterile
environment without a hint of a fungal network (we will consider
later how street trees, for example, can manage without a mycelial
network), and the second is a more general assessment of a very old
tree, Methuselah.

Methuselah, the famous California bristlecone pine thought to be
4,789 years old, didn't get to that age by playing fast and loose with
potentially pathogenic organisms. The tree, not unlike rock in appearance, lives among the rocks of the Inyo Mountains Wilderness,
the dry, cold range of mountains inland from the Sierra Nevada. The
height and width of the Sierra Nevada block the Inyo Mountains
from the warmth and damp of the sea, so the trees there must form
careful associations to persuade fungi to degrade organic matter and
harvest nitrogen from a long way off in the nutrient-poor ground of
the high forest where it grows. As a result, a delicate negotiation takes
place. A strictly regulated amount of sugar from the pine's photosynthesis is eked out to an associated fungus, and if the nutrient needed
in exchange is too far away for the tree's roots another fungus, at
arm's length, will harvest it and send it via long hyphae for the plant

to use. The tree has effectually built a power pyramid as exploitative as that of the early Egyptian kings. It controls the system via transport proteins, tuned over millions of years of trial and extinction to mete out the right carbon reward to the right fungus.

These are ectomycorrhizal (ECM) fungi: fungi that fear to penetrate their hosts' cell walls and instead form an interface called a Hartig net, a lattice of interlinking ectomycorrhiza that grow round the root and into the space in between root cells.[3] The hyphae release nutrients into the space around the roots, the roots take them up and use them for essential processes in the heart of the cell. In return, as much as 20 percent of the carbon captured by the tree can go to a fungal partner. It tends to be slow-growing trees from colder climates that negotiate to keep fungi outside their cells and form the ECM interaction. The fungi form a mantle around the root protecting it, and these interactions can continue around tree roots that are as much as 13 feet deep.

ECM fungi include the ceps and the truffles, the chanterelles and the matsutake. Their marketplace-like interactions with trees are widespread across temperate climates. In the case of bristlecone pines like Methuselah, however, even these loose associations may not always form. Scientists were intrigued by the way limber pines (*Pinus flexilis*) were jumping ahead of bristlecone pines (*Pinus longaeva*) up a slope in the White Mountains of California and investigated their interactions with fungi. The limber pines seemed to be getting ahead on these bare slopes because they quickly formed ECM associations with up to fifteen different types of fungi. Individual limber pines were twice as likely to form associations with fungi, and three times more likely to form extensive interactions than bristlecone pines. This seems to provide them with an advantage, including access to more nutrients, allowing them to grow faster. Meanwhile, the bristlecone pines continued to grow, just much more slowly, when supported by no mycelia at all. Is the secret of their slow growth and immense longevity an extremely picky attitude toward forming even guarded associations with fungi?

It may be possible for bristlecone pines to take or leave symbiotic relationships with fungi, but most trees will benefit from shaping

fungi to serve them in a more or less intimate relationship.* On the
damper soil of the Sierra Nevada, further to the west, the towering
Sequoiadendron trees of Yosemite National Park and the groves farther
south have formed a different kind of negotiation with their fungal
helpers: an arbuscular mycorrhizal (AM) network.[4] Giant sequoias
are very long-lived, but primarily they are the most massive trees in
the world and different from the sugar pine and beech that grow with
them in their groves that, like Methuselah, form ECM interactions
with fungi. With giant sequoias, fungi, notably those of the genus
Glomus, enter deep into their roots as they grow. The fungi develop
best on soil that has been burned, but will establish themselves on
trees over many years, whether burned or not.[5] *Glomus claroideum*
penetrates deep into the plant and forms arbuscules, shapes that look
like a little tree, in the cortical cells of the root. Unlike the ECM interactions that Methuselah might choose to form (keeping the fungal
hyphae outside the root cells but entering into a close relationship
with it), the giant sequoias are allowing the fungi right inside their
cells, almost becoming a hybrid organism with their fungal partners.

Under a fluorescent microscope the formation of an arbuscular
mycorrhiza in a tree root is a sensationally beautiful process. The fungal tubes, the hyphae, move toward the root and pool on it, forming a little foot, a hyphopodium. This structure then sends out more
little hyphae that force their way into the root and spread between
the cells, pouring into adjacent cells and forming a neat row of the
tree-shaped arbuscules, until the nucleus and vacuoles look like futuristic tree houses on a tree-lined street, pods sitting in the branches
of the cushioning fungus. If you are lucky you can see what happens
next. The fungus deregulates and unfolds the genetic material in the
nucleus, expressing genes that allow for free flow between carbon
compounds made by the tree and nitrogen and phosphorus, as well as
other metal ions, lithium, calcium, and magnesium. The trees are allowing the fungus to manipulate the very DNA inside their nucleus.
In return, they get a willing agent.[6]

---

* It is predicted that 90 percent of trees will form an association with fungi in one
way or another, but this figure has of course not yet been properly tested.

Hygrocybe mushrooms are very common under and on *Sequoia-dendron giganteum*.* For a long time these fungi were thought to be only saprotrophic and exist on dead wood, but it's now known that they also sometimes acquire nitrogen from strange sources high up the food chain and may, like the similar-looking *Laccaria laccata*, eat worms, subsequently transferring the nitrogen to the giant sequoias. The fruiting bodies of *Hygrocybe* are the waxcaps, brightly colored mushrooms with wide-set gills normally known as mushrooms of grasslands; their host associations are extremely flexible. Others are more specific, and one of the joys of mushroom hunting through a wood in the autumn is that the mushrooms can be easily associated with the trees they grow on. The porcelain fungus, for example, a delicate white translucent mushroom with a slimy top, nearly always grows out of the black, smooth bark of dead and rotting beechwood and so can be seen from far off. Once it has been deslimed it has a good taste, the umami coming from the glutamate it is packed with. In year on year of searching for it in the Devon wood where I grew up, I only found it twice: once in a dead branch of my favorite climbing beech, and once on a dead but still standing beech by the hedge boundary. I never saw porcelain mushrooms on the really ancient beech, hollow and rotting for thirty years, above the tiny old quarry where stone had been mined for the house. It was crowded with massive bracket fungi, *Ganoderma*, big enough to sit on until I was twelve or so. I always wondered whether the *Ganoderma* had kept the delicate porcelain cap away through some form of chemical attack and what it had taken the tree to keep the porcelain cap on the dead wood alone.[†] Was the great subterranean malice of that delicate white cap probing at the tree's defenses, trying to get into the live wood and digest it? Years later, in the Christmas after university, a branch fell off the climbing beech and we cut it up for the fire. There, in the

---

* *Hygrocybe* are endophytes; they will live in the bark or roots of *Sequoiadendron giganteum* but not form AM or ECM interactions.
[†] In this it seems I had it the wrong way round. Porcelain mushroom, *Oudemansiella mucida*, releases a powerful fungicide that deters/annihilates competitors. This was why it was alone on its branch.

pale-grained wood, were great threads of black compartmentalization, beautiful and dramatic swathes like an achromatic aurora.

The year the porcelain caps appeared was also the year my cousin
and I decided to poison our older siblings. Liberty caps and panther caps had appeared in the wood. A destroying angel, white and
white-gilled, was in the field just beyond and, thanks to another great
Roger Phillips book, *Mushrooms*, we knew exactly how poisonous
they were. We dismissed the panther cap, another beech associate,
with its beautiful brown cap and crusty white spots, as too dangerous and concentrated on the liberty caps, so small and meek and
gray and emphasized by the one-word entry: Edible? They were, of
course, magic mushrooms, *Psilocybe semilanceata*, the use of which was
banned in 2005, and luckily the stew looked so unappetizing that no
self-respecting fifteen-year-old would ever have touched it. If they
had, they might have seen inside a world in which trees and fungi
merge and coevolve and comanipulate—the world of strigolactones.[7]

Strigolactones are plant hormones, chemicals that trees exude to
control and shape fungi, and how they do this is in fact two stories,
the second a mirror image of the first in a dramatically concertinaed
time period. The first story is the development, over around 400 million years, possibly longer, of plant hormones that enabled plants
to balance the growth of their overstory and their understory, their
shoots and their roots. These hormones guided fungi to the roots,
with their supply of sugars and complex chemicals, and the fungi
were in turn coopted by the roots and used to gather nitrogen and
phosphate for the tree. Trees lacking in nitrogen and phosphate, even
sulfur,[8] started to exude strigolactones from their roots in order to
recruit fungi.* Finally, plant parasites evolved that, like fungi, sensed
the presence of tree roots through these hormones and entered them,

---

*Non-host plants of AM fungi do not upregulate strigolactones in root exudates
in response to nitrogen—and phosphorus—deficiency, but do upregulate strigolactones internally.

but unlike fungi took too much and gave nothing back. As a result, trees evolved in which the chemical structure of the hormone they used was very slightly different. These cultivars could escape the parasites.

The mirror image story of what humans know about these hormones starts in 1966, when scientists realized that some cultivars could escape the *Striga* or witchweed that was decimating crops like maize or sorghum in some parts of sub-Saharan Africa and isolated the chemical that appeared to be responsible, calling it strigol. In 2005 the role of these chemicals as an attractant and social lubricant between fungi and plants was discovered,[9] but it was only in 2008 that scientists from the Berkeley lab realized that strigolactones were a very important class of tree hormones, involved in cross-talking with the growth hormones called auxins and controlling all forms of branching.

Strigolactones are closely meshed rings of carbon that are a spin-off of the carotenoids, the carbon rings that give plants their orange or red colors. There's a wonderful carelessness in responses designed by evolution, responses that are the opposite of teleological. In this case, when plants are short of nitrogen, and particularly when they are short of phosphorus,[10] the carotenoid production pathway spins off, reacting to make strigolactones. In plants, the role of these hormones is to manage the balance of shoot and root branching so the plant does not outgrow its strength above ground. Over time this clearly acted as a signal to fungi that the plant would welcome, or at any rate tolerate, a partner below ground that would act as a potential nitrogen and phosphorus source, which is precisely what the fungi provide. Parasitic plants like the *Orobanche*, more worryingly, detect the scent of the hormones and use it as a signal for their seeds to germinate, after which the seedling will latch onto the roots of the strigolactone generator, which presumably is weaker and less likely to be able to throw them off. This is used by the truffle hunters of Syria, who follow the pale orchid-like flower spikes of the *Orobanche* parasite-flowers to try to find the *bint al ra'ad* or daughters of thunder,[11] as the desert truffles are called.

Trees are masters of supply and demand. Their supply chains are as robustly maintained as Napoleon's and, as with Napoleon's assault on Moscow, if the supply chains can't keep up or in some way get cut off, the result is disaster. Napoleon, however, only had to depend on supply chains up to the front. Trees have to balance the nutrients from the roots with the sugar from the shoots and so are dependent on a two-way traffic of information. They manage this by directing growth of both root and shoot with hormones, and this makes plants vulnerable to attack, because hormones leach into the soil and act as a signal to predatory fungi. But just as plants will switch between different subgroups of strigolactone to avoid *Striga*, so plants can adapt to avoid the fungi they don't want, while continuing to encourage the ones they do.

The flexibility of plant genetics allows different cultivars to minutely edit their strigolactone types, so that the fungal receptors blindly probing the soil to try to find them cannot recognize their usual partner, and the trees are able to select for fungal partners who will be more cooperative. Biochemistry allows this to happen in a wonderful way. The receptor is a protein that is folded into a certain shape, like a spanner; the strigolactone is a little ring like a nut that fits into the spanner. The result is that any distortion of the ring changes its ability to fit the protein. Over time, trees have initiated this structure so that the connection is made highly specific, allowing a tree to escape from any interaction with a fungus that might not be to its advantage.*

Strigolactones can also signal to nitrogen-fixing bacteria that the tree is ready to enter into a symbiotic relationship, but if the strigolactone changes, there will be no corresponding change in the bacteria. There will, however, be a change in the fungal composition of a rhizosphere,[12] and sometimes this can lead to further speciation of fungi, the creation of a new species that can respond to a plant's specific chemical variant. Strigolactones will change how and

---

*In the absence of strigolactones, hyphopodia do not form and interactions between root and fungus are drastically reduced; the signal transduction that strigolactones mediate is also necessary though the mechanism for this remains a mystery.

how deep in the soil a root branches, when and where a fungal spore germinates, and how one hypha branches with another. In this way, trees shaped the amazing diversity of fungi through delicate manipulation with strigolactones.

It's in the context of all of this that the idea of the wood wide web has germinated: an underground network of mycelia that links trees together almost like a world of underground inosculating roots. Humans have pattern receptors on the surface of their cells that will accept or reject another cell as self- or non-self; an example would be the antigens and receptors described by blood types. Plants don't have an equivalent, but are much more likely to respond to a surface in terms of how congenial or otherwise it may be. This is why trees will grow happily over stone, swallow up barbed-wire fences, or inosculate trees of different species. Fungi are more like animals and have a highly regulated mechanism of self-recognition, reaching out gingerly to unknown hyphae they come into contact with. They will form connections with hyphal networks of other fungi by means of anastomosis, creating a sort of synapse, and thereby creating potentially huge numbers of connections below ground. In doing so, they have the potential to connect all the trees those fungi are connected to.[13]

Over the three days of its life, the amethyst deceiver fades from a zingy upbeat violet to the deep-purple bruise of a furious storm cloud. It grows in deciduous and coniferous woodland and is widespread around the world. You can stumble across it on Hampstead Heath in London, in the Sam Houston National Forest in Texas, and on Mount Fuji in Japan. I went to look for it in Highgate Wood on a blustery day in October and found a cluster of tiny amethyst caps under an old beech. Even in the gloom of that afternoon under the beeches they were sparkling slightly, wet and fragile and pearled with water, translucent and otherworldly, or perhaps underworldly. They grew up out of the fallen beech leaves like symbols of umbrous perfection, round capped and delicately stalked, and they had the unsettling, clandestine look of underground things. I searched around for more, but they seemed to be alone. Apart from this one small patch of purple, the covering of dead leaves under the beech trees was unbroken, apparently innocent of any other life.

If you scrape away the drier top layer of leaves and pick up two well-rotted beech leaves that seem glued together, peeling them apart will often reveal the sticky mycelial threads of fungi. A little further below that, and in the earth beyond is a huge mass of life: not just mycelium of the amethyst deceiver, but threads of all the other mushrooms that might appear there briefly in the course of the year—morel and beechwood sickener and blue roundhead—as well as all the fungi that have no fruiting bodies at all. The most common ectomycorrhizal partner is *Cenococcum geophilum,* the earth-loving mushroom, which has no common name because it never makes a fruiting body and appears above the earth, if at all, as an insignificant black crust. Under the caps stretches out the network of hyphae, clustering in and around and entangled with the roots of the beech, probably slipping deep under the roots below the huge trunk, stretching further than the wide canopy, perhaps as far as the next huge trunk, and the next. It was hard to imagine that this network of huge gray tree trunks, glistening massively like an elephant in musth, might rely on an unseen, threadlike attachment to these tiny delicate caps, but we are not used to life hanging by a thread.

The concept of the wood wide web proposed a new and inverted world. Beneath every wood and forest there was a mycelial-mediated transfer of nitrogen and phosphate and sugar, signaling molecules, calcium and magnesium, glutamate and fats and amino acids, perhaps even electrical signals—a constant thrum of interaction from fungus to fungus, all essential for the life of trees. The first proof of plants transferring sugars to one another through fungal connections came in 1984 from the lab of David Read and his colleagues, in an ingenious experiment using radioactivity. Plants with or without mycorrhizal fungi were grown next to one another, and one in the pair was fed on carbon dioxide that contained the radioactive carbon 14. The carbon dioxide was taken into the leaf and photosynthesized; the carbon was locked down into sugar. After two weeks the root system of the other plant was then held up in front of radiographic film. In each case, plants with mycorrhizal connections showed evidence of the radioactive carbon in the plant it had been exposed to, the mycorrhizal network, and the other plant. Plants without fungal partners had not transferred the radioactivity between them.

So far, these findings were fascinating but couldn't be broadly applied. In the 1990s, Suzanne Simard conducted a similar experiment in the forest research center in Kamloops in British Columbia. She exposed pairs of tree seedlings growing in the forest to the same radioactive carbon dioxide and then measured transfer of radioactive carbon between trees. She found that as much as 6 percent of the radioactive carbon dioxide taken in through the stomata of a birch would appear in a Douglas fir some distance away, a level of exchange that she viewed as meaningful. After shading the firs and essentially starving them, they received more carbon from the unshaded birch than they did when they were happily photosynthesizing. In winter, when the birch had no leaves, the sugar flowed the opposite way. Here, Simard claimed later, was proof of a sink-source relationship: the trees were supporting each other by providing their excess sugar to one another in times of need. It was an example of mutual support in a healthy ecosystem: those with an excess supporting those in a time of need.

To anyone who has spent time in a mixed old-growth forest this idea is not hard to intuit. The moss piles up across the living tree, the *Sequoiadendron* inosculates the red cedar, the lichen half-clothes the leaf. Just as the cells in our body cooperate despite their differences, so there is a richness to the diversity in an old forest that implies a connected system based on a series of finely balanced feedback loops—and where better for this to happen than in the dark underground? It was also beautiful field science, taking place among all the excitements of bears and distant equipment supplies and mucky, interreliant ecosystems. The paper was accepted by the seminal scientific journal *Nature*, and with her coauthors David Perry and Melanie Jones, Simard changed how we look at the way trees grow in a forest.

The wood wide web was perfect branding and an immensely

compelling idea. Since the scientific revolution of the seventeenth century, the aim of science had been to separate an element of the natural world from its environment and learn the truth about it in isolation. This had been very productive, allowing chemists to isolate oxygen; Linnaeus to classify animals, plants, and fungi into measurable species; and botanical illustrators like Bauer, Ehret, and Catesby to paint one plant out of a forest ecosystem with absolute precision. It had led to Darwin's theory of evolution and the elucidation of the structure of DNA. By the 1980s, however, scientists and non-scientists had started to tire of the idea of constant dissection. What was the point of knowing the names of all the species in a forest if you still couldn't understand how the forest was functioning as a whole? By the 1980s ecology had gone a long way to reintegrate the pieces, but the idea of the wood wide web cemented that reintegration. It was a triumph of network theory and social science and fitted in well with other branches of intellectual thought that were also trying to reintegrate knowledge and remove the idea of the individual. The equivalent in history might be concentrating on the entire society of Macedon rather than the character of Alexander the Great. What if the most important thing about nature was not the individual tree, but the interaction of many trees through their fungal partners?

The problem was that, while the evidence for mycorrhizal connections with trees was strong, the evidence for trees connecting through those networks was, and remains, limited and inconclusive. Connections are messy, and scientists "have become vectors for unsubstantiated claims," says Justine Karst, a professor of mycology who works on mycorrhizal networks.[14] A paper published by Karst and one of the coauthors of Simard's earliest study, Melanie Jones, found that not only was there confirmation bias in subsequent publications, with studies showing evidence of mycorrhizal interactions more likely to be accepted by journals than those that did not show them, but the earliest evidence had also been overstated. Sometimes there appeared to be considerable interchange between trees, sometimes not.[15] In 2009 a genetic study that took samples of fungal DNA from the rhizospheres of different trees and showed the prevalence of different fungal types was often subsequently referenced as a paper showing nutrient exchange, when in fact it had been a purely genetic experiment. "Among peer-reviewed papers published in 2022," said Karst and her colleagues, "fewer than half the statements made about the original field studies could be considered accurate." The idea that the trees may have networks that stretch for miles or that the relationships are exclusively or even mainly cooperative seems to be simply wrong.

Most of all, Karst and colleagues objected to the idea that "mother trees" were supporting their offspring. What the evidence showed was that shared mycelium networks between older trees and their offspring sometimes retarded the growth of their offspring, sometimes had no effect, and sometimes helped the younger tree. In addition, Suzanne Simard's bestselling book on the subject, *The Mother Tree*, anthropomorphized trees and thus massively antagonized dendrologists. To start off with, the trees she was referring to weren't female, they were hermaphrodite. Secondly, her use of the internet or pipes as an analogy turned mycorrhizal networks into non-discriminating, empty pipelines, a picture that is very far from the truth of an active mycelium network. Most of all, anthropomorphized trees started to behave in strange ways. They started to talk and love and cooperate, to "choose" their offspring and display caring behavior. All these were human traits it was easy to cut and paste onto plants, but the result was a very human consistency, whereas a tree's ability to be dispassionate—to think and feel like wood—is one of its greatest wonders. Putting a nurturing mammalian face onto the giants of the forest was also a massive betrayal of the complexities of an organism that could be thousands of years old. Thinking of the 5,000 years in which Methuselah has had to negotiate existence makes simple narratives about the gentle exchange of nourishing sugars seem astoundingly trite.

The poetry of the wood wide web may have outstripped the scientific facts, but there remains a lot going on in shared mycorrhizal networks. It takes about thirty-five minutes for a hyphal tip to merge with another hypha in the lab and start to exchange particles, including vacuoles, mitochondria, nuclei, and fat droplets.[16] Other chemicals sped along through mycorrhizal networks may include poisons. In one experiment, juglones released from the leaves of a walnut tree traveled through mycorrhizal networks to accumulate round the roots of tomato plants and reduce their growth.[17] Volatile organic compounds, like the ones that seed clouds and can be used to prime trees to produce secondary metabolites, can connect across a system of mycelia and warn neighboring broad beans about impending aphids.

There is also a possibility that electrochemical impulses may be shared through mycelial networks. In the last few years green fluorescent protein has been used to visualize waves of electrical actions shivering across plant leaves in response to damage, particularly glutamate-connected damage. It demonstrates intuitively how reactive plants are behind their apparently immobile surface. In the messy, sticky, interrelated world of biochemistry there is high functional redundancy; the same chemicals are used over and over again for different purposes. Glutamate is the perfect example of this, the anion of an essential amino acid, glutamic acid, which is used by bacteria, plants, and fungi as a fundamental building block of proteins. If you tap "glutamate" into Google, Wikipedia will tell you it is the most abundant excitory neurotransmitter in the vertebrate nervous system. What it won't tell you is that glutamate is fundamental in pollen and spore germination* and found in extremely high quantities in mushrooms, where it is responsible for the umami taste that makes them so irresistible.

In the first *Avatar* film the scientist Dr. Grace Augustine explains the magic of the mythical planet Pandora in terms of connectivity:

> What we think we know—is that there's some kind of electrochemical communication between the roots of the trees. Like the synapses between neurons. Each tree has ten to the fourth connections to the trees around it, and there are ten to the twelfth trees on Pandora . . .

---

*If a tree falls in the forest does anyone notice? If a tree screams is there anything to hear? Like an octopus cradling a hurt arm, does this represent stimulation in any other way?

That electrochemical connection, glowing only a little less brightly, can be seen in the real forests of planet earth.

In 2019 an ambitious paper was published in *Nature* that mapped how fungi, bacteria, and trees collaborate on a worldwide scale.[18] The maps provide a new global projection of different interactions between fungi and trees underground, the individual components, if you like, of the wood wide web. Arbuscular mycorrhizal fungi, first recorded inside tree roots of the early Devonian period, 419.2 million years ago, are concentrated in a broad swathe round the equator, passing through the Amazon Basin, totally dominating Angola and the Democratic Republic of Congo, and sweeping round the Philippines and Manila. Ectomycorrhizal fungi, on the other hand, ring North America, associate with the majority (though not all) of the trees in California, go past New York, have an almost complete monopoly in London, Paris, Oslo, and Poland, cross easily into Russia, and continue past Moscow, beyond the Urals, through Siberia, and all the way to Vladivostok.*

Trees choose their fungal partners carefully, only feeding those they feel will not be a threat. Applying exudates, the unique mix of chemicals from a specific tree species, to fungal mixes has shown that certain trees chemically "ramp up" to attract certain fungi, and only some of these chemicals are strigolactones.

So, what's going on? Why would trees allow fungi inside their roots in the tropics but not in colder climates? The answer is probably energy and how much of it a tree can afford to lose. In the tropics, a successful tree will be energetically comfortable, as there tends to be plenty of water and lots of light. If it does well, it will most likely do *very* well. With huge numbers of species in a small area, competition is very high, and the hot, wet conditions encourage bacteria and other fungi to rot plant material down very

---

*Ectomycorrhizal fungi, which stay outside the roots, supporting them like part of the soil, include mushrooms like truffles and the fungi that feed Methuselah.

quickly, creating plentiful nitrogen that the tree can take up by itself from the upper layer of the soil. Phosphorus, on the other hand, which is essential for getting energy around the tree and allowing it to function fast, is at a premium, as there are so many other trees that need it. The result is that trees in tropical rainforests work on a boom-and-bust model. It's worth them taking the risk of internalizing AM fungi that may turn against them, because the same fungi will provide plentiful phosphorus, allowing them to race to the sky and reproduce fast. As a result, AM fungi are favored when plant growth is prioritized and nutrients are not limiting.*

Ectomycorrhizal fungi, by contrast, grow under drizzling temperate skies or in dry Mediterranean climates where energy itself may be the limiting factor and trees may have to grow very slowly, with little spare sugar left over to give. There's less competition, so trees are more circumspect about allowing a potentially pathogenic fungus into their roots. But the climate also means that organic matter takes longer to break down, and it's useful to have an efficient rotter—a saprotroph—on your side. ECM fungi are much more diverse than their AM partners, with as many as 20,000 species, and the oldest-known ECM fossils were found in pine roots preserved in the Princeton chert of Canada, only dating from 49 million years ago.[19] The theory—not completely proven but very convincing—is that when asteroid-induced darkness covered the earth and stopped photosynthesis for two years 66 million years ago, the saprotrophic fungi went on a rampage, and when trees, particularly the angiosperms, came out of the other side, they wanted to keep the saprotrophic fungi closer.†

Today, radical chemistry is one of the largest sources of carbon emissions, releasing an estimated 85 billion tons of carbon to the at-

---

*AM fungi are also ubiquitous among the gymnosperms, which seal off their roots to prevent close associations with further fungi. Only pines appear to have managed to break away from this suffocating relationship.

†In fact, environmental and host factors explain most of the variation in ectomycorrhizal diversity that can be seen in temperate woodlands across the globe. https://nph.onlinelibrary.wiley.com/doi/10.1111/nph.17892

mosphere every year (versus 10 billion tons emitted by the combustion of fossil fuels). It's a process known as enzymatic combustion, and 66 million years ago it was one of the only ways to obtain energy.[20]

ECM fungi evolved out of these saprotrophs. How this happened has been mapped out in the *Russula*, ECM fungi named after their role in Russian cuisine, where the mushrooms of even the most unpalatable species—*Russula emetica*, the fiery sickener, or *Russula xerampelina*, the crab brittlegill—are pickled and enjoyed. The Russulaceae are a large family, and mapping back their similarities and differences to their nearest saprotrophic relation shows that as angiosperms diversified, the *Russula* diversified along with them, often losing their saprotrophic enzymes. So how did plants manage to tame the fungi that were eating them into becoming their constructive partners? Through literally keeping them sweet. Feeding the fungi on an easy supply of sugar caused them to deactivate some of the genes for rotting wood. They may not have lost them altogether—some species of ECM Russulaceae form mushrooms on well-decayed logs, associating with roots within the wood, a dual niche that requires POD ligninases retained from white-rot ancestry.[21] These adaptations may explain the enigma noted in the introduction as to why trees so often send roots into their own heartwood and reconsume their fungal-digested hearts?

The most prolific genera of fungi are those, like the agarics, the webcaps, and the brittlegills, that are linked to specific trees and dependent on securing a place next to tree roots. They live, on average, longer than saprotrophic fungi, nursing a relationship that allows them to flourish as the tree they live on flourishes. The vast number of these species probably reflects a vast number of ways to secure a niche next to the roots and fierce competition while doing so from a mix of bacteria and other fungi. These interactions have been balanced and directed for millions of years by strigolactones. Some plant species, such as poplars and eucalyptus, will form plant associations with both AM and ECM fungi. While few AM fungi are host specific, host preferences and selectivity are very important, and most AM roots have a suberized exodermis that forms a permeability barrier around plant-fungus interfaces in the cortex, protecting the

relationship and preventing unwanted fungi from entering or pumping in confusing chemicals.

It's a delicate balance, and one that humans frequently interrupt.
An example of this is in central Mexico, where mining companies are
trying to reforest an area by planting junipers, which dig tap roots
deep into their surroundings and are highly efficient stabilizers of
soil. In the disturbed soil the junipers grew in strong relationship
with arbuscular mycorrhiza, and this disturbed relationship leached
into the more established areas where the native oak woods, slow
growing and shallow rooted, had relationships with ECM fungi
only. The oaks suffered, with many dying, and the evidence seems
to be that the ECM fungi, starving outside the tree roots because of
competition with the AM of the junipers, became pathogenic and
forced their way in.[22]

The host–fungus interface may have evolved in part as a way to limit
cheating by tightly coupling the cost and benefit of exchange from
both partners. Fungi that need lots of energy from trees leave less
for the tree to grow quickly, while fungi that need less energy from
trees allow trees to put more into their own growth.[23] ECM fungi
don't just specialize to the species, they also specialize to the environment they're in. Oaks are widespread around the world, from Iraq
to America, from China to Mexico, partly because they're good at
working flexibly with the conditions they're presented with, rather
than being intractable. They're also extremely tough—one reason
you can find so many old and superb specimens in the UK. Another
reason is the fact that they've been grown in parks and therefore been
protected and remained little changed for hundreds of years. Consequently, a variety of specialized fungi have adapted to this, including
the spectacular many-zoned rosette (*Podoscypha multizonata*), of which
the UK has about 80 percent of the European population. The fruiting body of the many-zoned rosette looks like nothing more than an
ornamental cabbage. Frilly and rust red, it's an endearing sight under
an enormous oak and normally grows around 7 feet away from the

trunk. As such, it's one of the inner-circle of mycorrhizally associated fungi on these ancient oaks, along with the other oak specialist, the oak milkcap (*Lactarius quietus*), which hunkers up under the trunk to send up its pale, oily-smelling mushrooms.

Recently, however, in parks in Norfolk and Suffolk in particular, oaks have been drooping and dying of a mystery disease. In acute oak decline (AOD) mature trees develop weeping fissures called stem bleeds, their bark lifts off, they become more susceptible to attack by beetles, and may die within four to six years.[24] At a conference discussing this and other tree diseases in the UK I met Elena Vanguelova, a biogeochemist who had jumped ahead of the curve to try to work out what was going on and had studied mycelium networks of oak at a particularly granular and wide-ranging scale.

What she and Laura Martinez-Suz, along with other colleagues, had found was that most mature oaks have a mycelium network that consists of three concentric and widening spheres. These spheres make up three levels of mycelium attachment, each characterized by different fungi. Certain fungal species associated directly with the oak roots; these were safe but unadventurous. Associated with them in a slightly wider sphere were further-off fungi, a little more adventurous but also cooperative. A third ring of incredibly distant fungi (some as many as 100 yards away) were associated with the second ring. Broadly speaking, the most distant ring seemed to consist of fungi that had the potential to be more pathogenic, but which compensated by being adventurous, explorers and mavericks that were good at bringing rarer ions to the tree. In trees affected by AOD the most distant rings of fungus didn't seem to exist. The fungi were not that diverse and didn't stretch that far, and the farthest fungi had lost their love of adventure, their passion for a quest. The tree seemed to be weakened as a result, perhaps through a lack of micronutrients, a sort of oak scurvy. It seemed to be not the presence of any pathogen, but the absence of that third ring of helpful fungi as mycorrhizal partners that was weakening the tree and causing the mysterious disease.

The root cause of oaks losing their outer ring is a mystery, but often seems to be linked to high levels of nitrogen in the soil and air, mainly from fertilizer spread on fields. The question is whether this is killing the mycelium networks far away from the tree because they are more likely to be under a field that is regularly manured, or whether the tree no longer needs its fungal helpers because there is a lot of nitrogen around and as a result fails to supply the glucose payment necessary for exchange of these rare but essential ions. It may be that the problems start even sooner, and trees fail to recruit the most distant ring of fungal partners because they are in some way blunted by elevated nitrogen levels.

Whatever the reason, AOD is sadly not the only tree disease connected to misfunctioning fungi. The effect on fungi of living in symbiosis over a long time period with trees is strikingly similar to the effect of human domestication on animals like sheep and dogs. Just as most dog species have now been human companions for at least 30,000 years and have morphed from great wolfhounds to valiant but diminutive dachshunds, so fungi that often act as ectomycorrhiza become less aggressive and less opportunistic and are enslaved to the easy energy that trees provide. Chemically they become specialized and less able to make energy out of anything else, relying on bacteria, for example, to perform the first stages of nitrogen liberation or to extract iron from complex minerals. Domestication is a dangerous game, however, and sometimes goes wrong. Particularly as a result of climate change and global trade networks, trees are now bombarded with fungi and bacteria that don't understand the rules and can break through defenses to kill a tree in one season.

The saddest and best example of this in our generation is the fungus that causes ash dieback, *Hymenoscyphus fraxineus*. Moving fast from China, where a different species of ash tree uses the fungus as an ectomycorrhizal partner, the fungus colonized European ash trees and found it could consume, rather than collaborate with them. As the infection starts, spores enter the leaves through the stomata, and the fungus then spreads down the trunk, feeding off sugar, blocking water from traveling up the trunk, and then, as dead wood and dead leaves drop to the ground, feasting off the decaying matter before sending up fruiting bodies to flood the air with more spores. It's the equivalent of angry dogs, bored of Pedigree Chum, turning on the humans in Britain and killing them before crunching their bones.

An estimated 125 million ash trees in the UK have been infected since *Hymenoscyphus'* arrival in 2010, and this figure is only expected to increase.[25] At one point 90 percent of European ash were expected to die, but thankfully this predicted figure has been coming down. Water, if abundant, can give trees enough breathing space to respond to the fungus and recover. Unlike elm, ash species contain a lot of genetic diversity, particularly in the UK, where two major populations, one from the Iberian Peninsula and one from northern Europe, met and mingled. One hundred and ninety-two sites of variation have been identified in ash tree DNA that are thought to provide the trees with some resistance. Of these, sixty-two are associated with the trees' immune systems—complex chemicals that can destroy or compartmentalize out wood affected with the disease—and ash is certainly fighting back. I went with a colleague to see ash trees badly afflicted by ash dieback being felled to prevent them from falling on a footpath. The great trees were thoroughly dead at the top, with very few leaves and the loss of many of their smaller twigs giving them a scarecrow look, but when the huge trunk came down we could see the cut made by the chainsaw was still sap filled, not the dry dead wood of some trees with their black hearts, but damp and, amazingly, a strong rose pink. The tree was fighting hard in the only way it could, by pumping out chemicals that might stave off the fungus. But with so many trees dying and so much fungus around, I wonder whether with time the trees, accustomed to shaping fungi, will once again be able to twist them to their advantage?

When young trees of many different types are planted in an area where trees have existed for a long time—particularly an area that has been disturbed by humans—some of the trees will almost inevitably die, quickly and seemingly without cause. This is not the mangled bark of a squirrel attack or the slow decline of drought: this is a sudden, stealthy paralysis and death. In temperate climates the culprit will usually be honey fungus, the psychopath of the fungal world. Mushrooms will sprout at the base of the trunk, yellow and multistemmed, and under the dead bark telltale strands of black are surrounded by intense, chalky white. Many trees in the same area at the same time will not die of honey fungus, and this defensive pattern is a display of which tree has successfully managed to modify its strigolactones to avoid an interaction between the mycelium of honey fungus and their roots.

Some tree species can avoid honey fungus better than others. Honey fungus is present in almost any temperate area that has had trees for some time, so woodland trees like beech or oak will have developed to avoid it. The trees that do die tend to be the pioneer trees: trees that often grow on open land where trees have never existed before, like the rowan, the birch, and the lime, the trees of barer mountainsides. Ash, which used to be relatively good at dealing with honey fungus, has become extremely susceptible to it as a result of ash dieback.

What is notable, however, is that the susceptible genera are susceptible whether or not they are a mountain tree in their native environment. In a UK setting, the woodland whitebeam from Sweden is as susceptible as the mountain ash from Scotland. Both are sorbuses, part of the rose family, and both, evidently, have avoided honey fungus well enough in the past that they don't have any good defenses against it. No one seems to be sure whether this is because they are pioneer species that live fast and can afford to die young, or whether they haven't had the exposure to honey fungus that would make them tweak their strigolactone to something less obvious and accessible. The pioneer trees are easily infiltrated and then easily killed, as the honey fungus spreads through them. This is intriguing because it is its mode of action that makes honey fungus so lethal.

When honey fungus mycelia colonize a root, it initiates contact through strigolactones just like an ordinary fungus. Rather than interacting with just one area of the root, however, a segment of the tree's influence, it loops round the full circle of the soil underneath the tree, and pumps toxins into the conductive tissue under the bark at the base of the plant, essentially ringbarking it. Initially, water still flows up into the wood; it is only the supply of sugar downward that is syphoned off, and the result is that the wood becomes water soaked, liable to windfall and white rot, which releases lignin-digesting enzymes while the roots are starved and die. When the roots are no longer functioning, the wood becomes dry and brittle, and finally a fan of white fungal mycelium runs up the trunk, prying the bark away from the wood, girdling the tree, and pumping it full of toxins.

The final stage of the honey fungus life cycle is digestion and spread. Lignin and cellulose are digested by finely tuned enzymes, more accessible sugars and proteins are siphoned off and used, and the vibrant mustard of the honey fungus sprouts along the base or the trunk of the tree. Tough black bundles of mycelia, the rhizomorphs, run through the ground like pseudo-roots, searching for a new victim before the old one is exhausted. In cases where forests are weakened, as is happening with ash dieback, the strength of the fungus is such that it can kill and consume playing-fields' worth of trees.

This Ashmageddon is a rare example, but there are cases in which networks of honey fungus have spread for thousands of years across thousands of acres. The record holder, an *Armillaria* in Oregon, is spreading rapidly and has reached 2,385 acres. Unsurprisingly it's nicknamed "the humongous fungus." How can trees possibly survive something so powerful? As with all tree problems, their solution is compartmentalization, evolution, and playing to their strengths.

In May, I went to a place called Hemelrijk in Holland, where thirty years ago two people, one Slovenian, one Belgian, had bought a field and then planted flowers and trees from all over the world. Some of their most famous collections are from Yakushima, a small island off the coast of Japan that is wreathed with fog most of the time, but has big stands of old-growth forests filled with *Cryptomeria* and *Tsuga*, with *Stewartia* and *Trochodendron* and with rhododendrons growing on their trunks.* All day it rained, but in the evening the sun came

---

*Old-growth forests that remain include the ancient temperate rainforest of Yakushima, the Belarus-Poland border, the giant sequoia and coastal redwood forests.

out and illuminated the trees. They were everywhere. By a lake were rare walnuts from Kyrgyzstan and Japanese chestnuts. The striking spiky leaves of an oak tree, "Hemelrijk Silver," a cross of specially selected valonia oaks from the Zagros Mountains of Iran and Iraq spun white flame into a background of other oaks. Rhododendrons spread flashes of vivid red, soft pink, and glowing cream under a canopy of hornbeams, oaks, and beech. The air smelled of *Philadelphus*, the fresh scent of new leaves and the poignant smell of spring. It was impossible not to breathe deep and feel renewed; it was heaven on earth.

The translation of Hemelrijk from Dutch is Kingdom of Heaven, and like all kingdoms it was younger than it seemed; a mixture of the world's wealth in the rich deep earth of an agricultural field, and just the tall oak trees lining the road and a few copper beeches had existed when they started to plant their trees. In this forest, which fitted the landscape like a glove, were thirty-five species of oak (over 2,000 species of tree), with trees that would have diverged geographically and evolutionarily hundreds of millions of years ago. An oak from Japan, for example, would last have bred with the Californian live oaks 200 million years ago. And their mycelium networks, their ways of shaping animals, plants, and all the other elements, will be different, with 200 million years of biochemical difference locked away in their biochemical pathways, something that can now be read in their genes.

What is fascinating about this is that, with vanishingly few exceptions, all these trees can coexist. Flung back into one another's company, across time and space, into a sodden Dutch field, they build on each other's strength to become a pyramid of strength. Thrust into the proximity of a stranger, they may grow up and shade out those that grow around them, but they will never attack them outright, and they may feed them surreptitiously below the ground. At the very least, their root exudates will be contributing to the soil around them and the growth of these very distant cousins. Through conserved mechanisms these trees can cooperate to build a kingdom of heaven in a way that most animals could not. And if, to foolish human eyes, after thirty years they will look as though they have become a forest, with every year the interactions between the trees will become more and more complex, just like human communities. Both human and fungal ability to adapt and communicate across half a world of borders is structured by the trees that shaped our development. The difference is that our worlds are not infinitely adaptable, because our proteins, wretched lumpy things, can't just double and adapt themselves as trees can. So if we were taken from the cloud forest of Yakushima and transplanted into a damp field in Holland, we would sicken and die, just as so many Europeans who were colonizing tropical climates did.

(*Above*) Storks nesting in valonia oak trees growing along the Great Zab River in Iraqi Kurdistan. The rise of the Zagros Mountains led to the formation of different species that were then selected by humans. (*Below left*) Fossilized *Archaeopteris* root systems from 390 million years ago, photographed in the Cairo site, upstate New York. This particular root system stretched for 36 feet. (*Below right*) One of the ancient junipers of the Ziarat Juniper Forest on the Pakistan-Afghanistan border. Note the roots forced above the ground by the thin soil.

(*Above left*) A burl the size of a small car on a giant sequoia growing in the Trail of a Hundred Giants, Sequoia National Forest, California. If the redwood falls over, the tree can spring up again from the burl. Some have suggested that this evolved as a mechanism for regrowth when dinosaurs pushed trees over. Now they are more likely to respond to fire or chainsaws. (*Above right*) Giant sequoias commonly grow as twins, fusing root plates and occasionally linking vascular systems. (*Below*) A controlled burn in the Trail of a Hundred Giants.

(*Above left*) A horse chestnut growing on the cliffs of the Vikos Gorge, northern Greece. From this original *refugia* horse chestnuts have spread across the northern hemisphere. (*Above right*) A flower head, or candle, of a horse chestnut, showing the adaptations to attract bees (and, coincidentally, humans). The red patches indicate pollinated flowers and repel bees close-up by appearing black, sending them into the unpollinated (yellow) flowers, which have added UV sparkle in bees' eyes. (*Below left*) The pollen-producing catkins of a hazel (*Corylus avellana*) tree are optimized for wind dispersal. The female flower is mostly inside the bud, and only the sticky, bright red styles are visible. (*Below right*) The red flowers may also attract birds to serve as an insurance pollen vector; we can see this in the *Bombax ceiba* trees growing in Jallo Park, Lahore.

(*Above*) The Rio Negro near São Gabriel da Cachoeira. One of the major tributaries of the Amazon, the river's name comes from its deep brown color, which is the result of a high concentration of humic acid from the breakdown of the surrounding trees' leaf litter. (*Below left*) The white sand of the Serra do Curicuriari in Brazil is an old soil and very low in nutrients. Underneath the canopy layer there is an incredible diversity of legumes: trees which have pods and the ability to fix nitrogen. This particular pod, from a fairly small tree, was more than 6 inches long. (*Below right*) The plant immune system consists mainly of chemicals. These may be strongly antibacterial and antifungal, or simply toxic, and are often collected and used by humans. Baré incense, the response of a group of trees including breu (*Protium heptaphyllum*) to attack by beetles, pictured here next to the Rio Negro in March 2024.

(*Above*) Joshua trees (*Yucca brevifolia*) in Joshua Tree National Park, California. As monocots they don't lay down growth rings so their age is hard to assess, but many are over 150 years old. When their growing tips are damaged by frost they both flower and branch, producing strange fractal silhouettes. (*Below*) Cedars of Lebanon near Kahramanmaraş, Turkey.

(*Above*) Golden aspen, probably a single clone of *Populus davidiana*, suckering down a stream through fir forests in China. (*Below left*) The Chinese golden larch (*Larix potaninii*). The lichen that covers it (probably *Usnea*) and the red berberis growing underneath are some of the only plants able to survive in the highly acidic environment created by the tree. (*Below right*) The lacebark patterns of *Pinus gerardiana* growing in the Hindu Kush. It shares the flaky bark of *Pinus bungeana*, its closest relative, which grows across the Himalayas in Shanxi.

(*Above*) A date palm grove in Anbar province, Iraq, obscured by a dust storm. (*Below*) Yunnan: Bai Ma Shan, China, in autumn with the gymnosperm *Larix potaninii* and the angiosperm *Populus davidiana*.

(*Above*) A cloud forest in Sichuan: the foothills of the Gonggar Shan in summer. The pinnate leaf on the left is *Pterocarya macroptera* var. *insignis*; deciduous trees hang on to the wet cliff; and pines grow along the ridge. (*Below*) Rainstorms over the turbulent forest canopy of the Amazon near Manaus, as seen from the tower of the National Institute of Amazonian research. The currents of moisture are so strong that they are known as flying rivers and can be sucked thousands of miles in from the sea.

CHAPTER 6

Trees shaping plants

How trees shaped the plants underneath them, and how in some cases those plants grew up to become trees themselves

High up, red copihues (Lapageria rosea) *dangle like drops from the magic forest's arteries.*

Pablo Neruda, *Memoirs* (translated by Hardie St. Martin)

In the British Isles, bluebells are a sign of ancient woodland, but under gnarled ancient trees is not always where they flower best. One evening every May I go on a pilgrimage to a wood in Devon to see a hovering vision, a perfect haze of clear stalks above green collapsed leaves and heads of hard blue. The evening light shines through the tall oaks' canopy at just the right angle to outline each flower head in blue and gold. The scent is honey and smoke, and bees keep pollinating until the sun sets. You can tell from a lack of understory and a certain uniformity in the size of the trees that these oaks were all clear-felled about a hundred years ago and then the stools simply allowed to grow back, fast and all together.

The magic picture of dappled sun on blue tells its own story. Despite being a woodland flower, bluebells need a surprising amount of light. Their leaves appear in winter, and they flower at the time trees produce their earliest leaves, delicate and pale green, and they like to be damp. Trees will trap moisture under their canopy and give them a cool, moist environment, but they also favor a wet west wind off the Atlantic. This makes them remarkably fastidious about the places where they will grow. While they push through the bracken to blue the cliffs of Lundy Island, even under trees bluebells will not go south of Bordeaux or east of Brussels. They will sweep through the woods of Aberdeenshire and are at their best in woods in Kent and Sussex, but they also produce great drifts of blue along the coast of Ireland, in the karst landscape of the Burren.

Some plants are quite clearly tree adapted. Vines and epiphytes are the obvious ones, a *Pothos* spinning in circles to catch a trunk, a *Wisteria* hanging its gridelin flowers from the top of the tree's canopy, or a *Monstera* moving toward shade to find a big tree to climb. But if an iconic woodland plant like a bluebell can grow in the presence or absence of trees, what does this tell us about how trees have shaped plants? The writer Zoë Schlanger has posed a question about the different requirements of plants: "What if, instead of hunting, our food was sunlight that rained down on us, so we were bathed in it and we had to evolve only to be prepared to receive it?"[1]

To that question I want to add another. What if our food, the sunlight that rained down on us, was consistently blocked off by a canopy of plants bigger than us? What if most plants evolved in the equivalent of a very repressive feudal system?

Over millions of years of plant development this was the reality. Trees covered the earth, and any other plants—anything non-woody—eked out an existence underneath them or perched on their branches. We're so used to shady forests and dappled woodlands that we can easily forget what they mean. The green coolth of forest bathing means an almost complete blotting out of light and its life-giving energy from the ground, and therefore no carbon fixation—total starvation for plants. It took over 300 million years for a true multicanopied layer to develop in forests, but before that most plants would have had to live underneath canopies of huge conifers—a world of towering tree trunks and epiphytes with little undergrowth apart from mosses, ferns, cycads, and horsetails.

In fact, most of the plants we see around us today started to evolve from one common ancestor sometime between 167 and 199 million years ago. The vast majority of plants (more than 352,000 species) are angiosperms, and their ancestors, the first flowering plants, started to evolve as part of a warm, wet forest undergrowth, lying low under conifer canopies that were gradually losing diversity. Analyzing the genetic lineages of flowering plants and mapping them onto fossil evidence shows that around 100 million years ago the orders of angiosperms diverged as Gondwana broke up,★ but it took a great extinction event, annihilation by the Chicxulub asteroid (the one that killed off the dinosaurs), for them to become truly dominant, and come out from under the shadow of the trees. How they did so is a long story, but we can see a speeded-up microcosm of how trees shaped plants by looking, ironically, at the areas where trees have all but disappeared.

One such place is the Burren on the west coast of Ireland, a remarkable limestone landscape that spreads in rolling plateaus into the sea. It's a bleak littoral, battered by winds across the Atlantic, and huge waves that roll in unhindered by anything apart from the tiny outcrops of the Aran Islands for more than 4,000 miles across the Atlantic. Before the Bronze Age, this landscape was densely forested, but humans and climatic changes have cleared nearly all the woods, leaving only hazel copses, the occasional very small tree, and a tiny patch of Irish pine. Remarkably, however, a very large number of woodland plants remain; in fact, only three of the sixty-four woodland plants you would expect to see in the area have vanished along with the trees. Alongside the bluebells, wood sorrel is everywhere, and so is wood sage, growing successfully miles away from any tree. The rare pyramidal bugle, the hairier, squatter relation of its tall dark-blue woodland cousin, also thrives. The Burren is an unusual place, and many of its mysteries remain unsolved, but the presence of so many woodland species here, which came with the trees and were left stranded like driftwood after a high tide as the woods withdrew,

★ The orders of angiosperms diverged as Gondwana broke up and moved apart, and some of the most beautiful are the Proteaceae, a group of plants that include *Embothrium*, also known as the Chilean firebush because of its coloring, a tree wrapped in dark foliage with bright-red tubular flowers. Most of the family are found in South Africa, Australia, and South America, although others are known from Hawaii and Mexico, Madagascar and India.

is like an allegory of plant development. Even when it seems that plant species have always lived away from trees, we will find the influence of trees in their development. In most cases worldwide, as in the woodland plants of the Burren, plants have been shaped first of all by trees, developing under them and then adapting out from under them.

Most woodland plants thrive in a damp environment that remains cool in the summer and is not too cold in the winter. Trees provide this by acting as protection, erecting a million tiny walls to trap moisture, stop strong winds, and shelter the ground from frost. The Burren provides this through deep clefts in the limestone, some caused by water and some by tectonic movements. Sections of the limestone pavement are as flat as its name suggests; thin slabs of limestone split by tectonic movements and thousands of years' worth of water into regimented rows of razor-sharp rock. Glaciers scraped the rock clear of soil, and the pavement and its cracks are known as clints and grikes—ominous, hard, scraping words. *Burren* means "the stony place" in Gaelic (the word *barren* comes from the same root), and the hills that rise up from the sea appear from below as though they are pure rock, unsoftened by any living thing. This is an illusion: a bird's-eye view shows the natural terraces of the limestone lush with soil and plants. The terraces were formed by climate fluctuations that induced changes in the hardness of the rock, with the result that, looking up, you can see only the gray, stark surface of the steep cliffs, not the fresh green of the plateaus and hollows where soil and water could collect.

In this fresh greenness, as well as in all the cracks in the rock, an incredible wealth of plants can grow. This is a karst landscape, a system of dissolved limestone riddled with a secret system of caves and conduits and springs gushing suddenly out of the cliff, which allows a million niches, each differently filled. My favorite of these plants is the Irish eyebright (*Euphrasia salisburgensis* var. *hibernica*), an extraordinary plant with deep-puce leaves and rakish white flowers—a yellow eye and specks of deep purple—which is hemiparasitic on thyme. It was once used to cure conjunctivitis. This variety is now found only in western Ireland, Newfoundland, and Labrador, a dramatically disjunct distribution! Two flowers only distantly related, mountain avens (*Dryas octopetala*) and the beautiful hoary rock-rose (*Helianthe-mum oelandicum*), lurch across cracks and have been pressed by con-vergent evolution into the same shapes: dark-green leaves and open, yellow-centered flowers.

Less obviously exciting when I visited were two woodland flow-ers that I know from Devon woodlands: sanicle and wood anemone. Wood sanicle has deep-green leaves and flowers like tiny white, fluffy pompoms—in Devon it appeared one year after we had coppiced the hazel and was a beautiful, leggy interloper. Wood anemone, which some people called windflower, has delicate, wind-waved, white-blushed pink flowers—not showy but deeply beautiful, gentle and with a hint of yellow. On the Burren both of these were rooted deep in the cracks of the limestone. Perhaps even more strangely, Hi-bernian ivy, an Irish variation on the wildly successful ivy that grows from Japan to Spain, and honeysuckle were growing prostrate along the ground. A sickly beige *Orobanche*, the ivy broomrape, lived para-sitically on the ivy, and the yellow, tubular hooks of honeysuckle flowers carpeted the ground. The few trees that also grow in the cracks tend to adopt the same prostrate position. Purging buckthorn, juniper, and creeping willow are the trees you would expect to find in the area, rarely reaching above their sheltering limestone cracks, but a man I met in the Burren told me the story of the avocado, pre-sumably cast off from a picnic by a human imitation of megaflora, that grew happily for a few years, sheltered in its grike from the frosts and blasts of the coast. The limestone absorbs heat and gently trans-mits it again—it is a great heat sink and, along with the Gulf Stream, ensures that this area of the coast is warm and temperate, and in the past it was valued as winter grazing.

The crouching trees tell us that the wind has the whip hand in every situation. But what we also learn is how adaptable and resil-ient are the species shaped by trees. It is only competition for light that makes the honeysuckle and ivy grow upward—as soon as they can find a nutrient-poor area that allows them full sunlight they will grasp it. The limestone soil of the Burren is poor and thin, and the ability of these woodland plants to germinate in the dark, deep in the cracks of the grike, is what gives them the advantage over the other flora that might outcompete them. Most of all, they can run from grike to grike without ever going out onto the clint.

As a result, these plants have succeeded in hanging on where most trees have failed. I was on the Burren to visit a mysterious population of pine trees, a tiny pocket of survivors from the woodland that had nursed the plants of the Burren out across the bleak coastal limestone before being destroyed themselves. Most people believed that they had vanished entirely in 400 BC, replaced by Scots pine reintroduced in the seventeenth century, but in 2016 genetic studies showed that these few trees had survived from the last Ice Age.

Ireland is rich in peat, and pollen grains preserved in peat tell us about the history of vegetation in Ireland since the last Ice Age. What we learn is that Ireland was dominated for several thousand years by the vegetation of the Arctic tundra, of which the beautiful mountain avens, creamy and low-growing, is a prime example, found mainly around the Arctic Circle. About 10,000 years ago the climate warmed up and woodland quickly took over. Juniper was the pioneer tree, marching across the hill in a shrubby vanguard; mixed woodland of birch, hazel, and pine followed, and then about 8,000 years ago mixed elm and oak forest with yew and ash developed, leaving pine to survive on drier soils. The climate became wetter after that, and alder more abundant, while pine declined. About 5,900 years ago something, possibly disease or people, decimated the elm, and in the Neolithic period, about 5,500 years ago, people set to and cleared vast areas of the landscape. In the Burren, areas of rock that had trees lost them forever, while the ground, fertilized by seaweed, was intensively farmed in the patches where soil still existed.[2]

Where did Ireland's trees come from? And what influenced whether they came or went? How long had they been in particular areas? In 1651, one of Oliver Cromwell's generals, Edmund Ludlow, wrote that the Burren had "not enough water to drown a man, wood to hang one, nor earth enough to bury one."[3] He was wrong. As mentioned before, while in most areas Scots pine had become extinct by 400 BC and was reintroduced in the seventeenth century, in one area, deep in the Burren, there was a continuous record of its pollen from about 8,000 BC onward. This fits well with the idea that all trees had to migrate into Ireland about 10,000 years ago, after the glaciers retreated. Before this the whole of Ireland was covered in an ice sheet, shared with Britain: the British-Irish Ice Sheet. If any land existed next to the ice sheet it would have been extremely arid, and in permafrost, completely unlivable for any of the plants currently found in Ireland. So how did trees spread into Ireland as the ice sheets withdrew, and what clues does this hold for how trees shaped plants?

Astonishingly, the climate in Western Europe changed from Arctic to temperate in about a decade, and in Ireland the bare Irish tundra quickly became colonized by trees. Juniper, its berries spread by birds, was the first, swiftly superseded by birch, with its huge numbers of windblown seeds, which was then in turn replaced by hazel. In the environment created by all these species, woodland flowers also spread up into the land that had been under the ice. The ice had been thick— more than half a mile in some places—and sea levels were around 400 feet below their current level. As the ice melted, the difference in pressure between where ice still was and where ice had gone forced up the earth's crust, and the result was a moving land bridge across the Irish Sea from 11,000 years ago until about 9,600 years ago. Pollen shows trees moving northwards, covering the south to north of Ireland over a thousand years. Birch, elm, oak, and pine all moved north together, ducking through valleys, skirting round mountains, and, in the case of pine, clinging to the limestone. By the time lime trees arrived in Wales 7,000 years ago the land bridges were submerged, and as a result limes never spread into Ireland, although both they and beech trees flourished when they were introduced.

With trees came flowers and associated plants.* Ivy will die in temperatures lower than around 5°F, and honeysuckle will struggle,[4] so moving onto an island with still-fluctuating temperatures after the Ice Age was a challenge. However, tracking the movement of ivy and honeysuckle through the pollen record shows them coming in

---

* The earliest plants had arrived 13,000 years ago. They included the Arctic *Dryas octopetala*, a plant of the tundra, and other more quick-colonizing plants. https://www-jstor-org.lonlib.idm.oclc.org/stable/20728598

just after the juniper trees that had colonized the landscape peaked: honeysuckle almost 11,500 years ago, ivy 10,000 years ago. They moved in under the cover of trees, sheltering in the shadow of their branches.

When we look now across a Burren devoid of trees, it can be hard to gauge just how important they were. No one really knows what led the once open forest of Ireland to crash, but it seems to have been a combination of the spread of peat bog,[*] a warmer wetter environment, and the impact of humans. The first evidence for humans in Ireland is from a lacerated bear knee-bone radiocarbon-dated to 12,500 years ago,[†] so it is safe to imagine that humans spread into Ireland with the trees, and that they, like the flora, sheltered in their shadow. However, some of the next evidence of humans in the west of Ireland shows them destroying the very trees that protected them as they moved in.

Peat-cutters working in the bogs of Ireland in the nineteenth and twentieth centuries frequently found stumps of pine trees that had been cut down thousands of years before. A Mesolithic site indicates usage of huge amounts of pine: not the side branches or twigs but the timber—the trunks and main branches. Neolithic burial sites have been excavated to show that 85 percent of the wood used was pine, indicating a massive clear-cutting of the pine woods, which would not regrow. When a Mesolithic stone axe was used to cut down an Irish pine tree, everything that made the pine tree hang together broke down, and massive, catastrophic loss of water occurred. Unlike oak, pine cannot resprout and survive being coppiced: the tracheids that conduct water to the top of the tree are irreparably damaged and cannot mend or regrow.

The tree was doomed, and soon pine after pine rapidly disappeared from the landscape. Willow, alder, and hazel (and even yew!), on the other hand, can grow back once they're cut down; it's the reason they cling on in their dwarf forms on the Burren, and also why the honeysuckle and ivy, the sanicle, and even the bluebells survive.

---

[*] Like the destruction of the mammoth steppe, the spread of bog often heralds a massive change in climate and vegetation.

[†] Found in the Alice and Gwendoline Cave in County Clare.

To visit the only patch of Irish pine that survived we came down from the high Burren and had to climb over three stone walls, wade across a bog, and wriggle through a hedge. We could see the remaining trees by their darkness and, when closer, by the redness of their bark and the domes of their canopies. The pine population is down in a dip, open spaced, surrounded by grasses and sedges, roses and buckthorn and guelder rose, and growing out of the damp land around broken limestone. It looked like most other Scots pine, but its needles were a little sparser, and there was something different about its bark: it looked a little more, in fact, like the survivor it was. Young pines were coming up through the scrub around it; encouraged by the farmer who protects it, the population was clearly thriving. Around it grew the woodland flowers of the open Burren, but they were not happier under the trees; they were smaller, scarcer. In fact, it was clear that the trees, like an overprotective teacher, had only held the flowers back. And, counterintuitive though this is, at a profound level it is the story of almost all plants on a much deeper timescale. Even plants that grow best in the open were shaped by trees in the shade.

𝔻

Roses in the evening sun of walled gardens, poppies in red swathes through fields of wheat, massed sunflowers mapping the sun from east to west, even love-in-the-mist coyly sunning itself in a border: it's easy to think about flowers drenched in sun and forget about the links in the chain that enabled them to behave as they do.

The appearance and identity of the first flowering plants and the speed with which flowering plants came to dominate the world have long been exercising the minds of evolutionary scientists, including Charles Darwin. As recently as 2017, Christenhusz, Fay, and Chase wrote that "the precise way flowers developed, and from what pre-existing group of gymnosperms (seed plants without flowers) flowering plants developed, is still a mystery."[5] For many years scientists thought the magnolias, woody plants holding up great waxy flowers to lumbering beetles, were the closest surviving relatives of the earliest flowering plants, but recent studies have overturned this theory.

Slightly under 200 million years ago, while the dinosaurs still raged up and down with their heads in a canopy of conifers, a new plant evolved under the trees on the forest floor that would revolutionize the world. In 1999 a group of evolutionary scientists from different disciplines across Canada and the United States came together—the botanist, the geologist, the ecologist, and the integrative biologist— and, as well as looking at the plants in the wild, they created a garden in Alabama, filled with all the plants that are known to be most closely related to the common ancestor of all the flowering plants. If you map back the DNA of flowering plants you find that four groups of plants that we know today diverged early from the main tree of flowering species. These four groups still retain ancient characteristics and indicate that the earliest flowers liked water, disturbance, and shade. Of the earliest side branch just one species survives, a wavy-leaved scrambling shrub with small creamy flowers that now only grows in New Caledonia: *Amborella trichopoda*. The second group of the four are the water lilies, the Nymphaeaceae, with floating tranquil pads and waxy flowers open to the sun or, in the case of the giant Amazon water lily, *Victoria*, open at night. The third group are the Austrobaileyales, fragrant plants and climbers including *Illicium verum,* the plant that produces the spice star anise. Finally, there's the obscure and interesting group of the Chloranthaceae (literally "the green-flowers"), seventy-seven species of flowering plants that grow in China and Japan, Central America, Madagascar and Malaysia, and from the Marquesas to Borneo. All these have very simple flowers with plenty of exposed pollen to attract beetles, the earliest insect pollinators.

In their paper, "Dark and Disturbed: A New Image of Early Angiosperm Ecology," written after growing these four groups together and comparing them, the scientists postulated that early angiosperms developed in wet, turbulent, tropical forest. Developing under this oppressive shade of towering conifers forced the early flowering plants to adapt quickly to their environment—to have, in other words, vegetative plasticity. Initially this may have been a handicap, stopping them from forcing their way straight up toward the sun, but it evolved to their advantage. Imagine the kind of gloom you might find in a fir plantation, and the areas where this gloom lifts slightly: a bank of shaley soil, for example, or a stream or pond, or an area where a tree has fallen. These are the areas where a little more light will penetrate down to the forest floor. Light is carbon fixation, which is the food of growth and life, and if a limb falls off a tree it's like a free lunch: you need to move rapidly into the gap where it's fallen.

The flowering plants, usually called angiosperms to distinguish them from gymnosperms, have their seeds well-hidden in the central part of the flower, the ovary. Pollen needs to grow a tube along the style and into the ovary down which the male cell can move to fertilize the egg, which becomes the seed. The seeds remain in the closed ovary until they are ripe. Many angiosperms also developed flowers with petals and often scent to attract insects to move pollen from one flower to another, thus ensuring cross-fertilization, which is to be preferred to self-fertilization. Through the millions of years flowers and insects and even birds have evolved in tandem, often needing one another in subtle and intriguing ways.

The first way the flowering plants could colonize new ground was to produce small seeds that would enable them to establish themselves quickly on barer terrain that had been disturbed by animals or a tree falling down, or in open water—but not in deep leaf litter. Once that seed had landed somewhere and germinated, it went through a diversity of vegetative regeneration modes that allowed it to cope with disturbance. The new plants could sprout from their base at any point in their life cycle, and their seedlings could pass through a phase of creeping along the forest floor until they found somewhere to send up shoots. They had rhizomes that could be moved and still regenerate, and their roots and shoots could layer and root again, which made them extremely resistant to trampling by herbivores and other forms of disturbance. Rather than having one dominant leader shoot, the new plants had many possible shoots and could respond to

limbs falling off trees and temporary burial. All these are adaptations
to the understory of the forest floor, to life below trees. Sprouting, it
appears, is an ancestral trait of the flowering plants.

Underneath their conifer overlords, the new flowering plants
also needed to capture every bit of light possible, and they did
this by moving along rapidly shifting streambeds, over soil that
was too poor for trees to grow, or into areas where large animals
had cleared a patch of ground. All the earliest known flowering
plants have adaptations to do this, and counterintuitively these
arose from the production of living cells with thin wall-like tissue
called the parenchyma, which probably arose at first as an expen-
sive misdirection—possibly something even akin to a tumor cell in
humans—but soon became invaluable.

The ferns and the cycads and other non-flowering plants cope
with shade by reducing their metabolism to a minimum, growing
very little or not at all, and biding their time in the darkness.[6] It was
parenchyma that enabled the new flowering plants to capture more
light from the darkest shade. They allowed development of a new
layer in the leaf, the spongy parenchyma, which acted like a hall of
mirrors. Here the tiny amount of light that made it down through
the canopy of the conifers to the forest floor was trapped in a sort
of watery maze within the leaf. The light would bounce backward
and forward, scattered in all directions by a mass of light-water interfaces until it hit one of the photosynthesizing cells almost ran-
domly distributed through the layer. Unlike the earlier system, in
which a neat wall of aligned rectangular cells sat across the leaf, this
chaotic bouncing massively increased the length of the path that
photons of light took through the leaf, and therefore increased the
chance of a photon of light being absorbed and used.★ It's so simple
and so brilliant, but it took this lucky accident of evolution for it
to happen.

★ The only problem is that when there is very strong light or lots of it bouncing
around, the protein-metal complexes that absorb the light first of all can be damaged, and this again is seen as one reason that the flowering plants have spread as
they do.

Once established, the same concept—fewer regimented cells and, instead, diffuse and random parenchyma—could revolutionize the way stems and roots grew and repaired themselves. Think Genghis Khan's hordes rather than Alexander the Great's phalanx: a mass of cells that could rush where needed and direct themselves where necessary, rather than a rigid formation that, once broken, has a hard job filling in the gaps. The *Amborella*, ancestrally the oldest, still has tracheids up its trunk—the stiff, coniferous, lignified cells—but the water lilies are starting to lose the stiff arrangement of the tracheids and have some longer vessels where cells lengthen into tubes through which water can whiz up and down, while the star anise vines and *Chloranthus* have extensive vessels supported by fibers and parenchyma. The result of this adaptation is that the energy cost of putting up a shoot and supporting a broader, veined leaf area is considerably lower in the fiber-supported vessel stem than the carbon required to put a layer of lignin around every tracheid cell. It's rapid, inexact construction, but it's quick and it works. And it was this that enabled further changes.

Reticulate leaf venation, a netting of veins across a leaf, allowed the development of larger, more flexible leaves that could catch the flecked and dappled light of the understory, and the angiosperms started to thrive. Creeping along the ground or through the trees, with many meristems from which they can reproduce, was the perfect way to get a foothold in a crowded Mesozoic forest frequented by dinosaurs. The angiosperms lurked in the darkness and the wet in various different forms before branching out into the sun: as the now extinct *Ficophyllum*, a climber with large leaves like a fig tree, as the green-flowered *Chloranthus*, or as the forerunners of the star anise.

In the sun, the new adaptations became yet another form of advantage. A leaf is designed to capture sunlight, which means it can very easily overheat or suffer from photoinhibition. The new arrangement of water-carrying vessels allowed plants to transpire so fast that they could cool off in the sun very easily. Earlier plants had worked around this by limiting the rate at which they grew and photosynthesized—you can see some of these adaptations in the gray-green, wiry, leafless stems of *Ephedra* (its pollen known in the late Triassic) of southern Afghanistan and semi-arid climates across the world. But the new plants had no need to do this: they could photosynthesize faster than ever. This allowed flowering plants to grow in places that would have been slow or inconceivable to master previously. The poplars of the Gobi Desert are a perfect example, shaking the heat off their leaves by transpiring fast with the water obtained from the deep euphratic layer, a process known as hydraulic redistribution.[7] Other examples include the gentians of the Alps growing quickly on the glacial meltwaters or the proteas of South Africa, photosynthesizing fast under burning sun.

Many tropical species that today thrive under the canopy of the rainforest display similar properties to those of the earliest angiosperms: remarkable plasticity and the ability to grow, change, and adapt to many different stimuli under the canopy. The earliest flowering plants, developing these tricks under difficult conditions, could take the skills learned in the forest and apply them elsewhere to great advantage. Like children growing up in tricky circumstances, their need to adapt fast allowed them to do great things. The strangeness of these clues from deep time was driven home one day when I was walking along the beach on the Sussex coast. When we were children my grandmother had occasionally come home triumphantly with a fossilized shark's tooth, so I was watching the water wash back and forth and was surprised to set eyes on what looked like fossilized star anise. Like hippo's teeth under Trafalgar Square, it seemed utterly anachronistic, on a cold, windswept April day in England, to be looking at a spice grown in China and Japan rolling around in the little waves at the edge of the shore. But these fossils were just that: seeds of angiosperms, 46 million years old but maintaining the shape of the earliest angiosperms, reminders of a time when tropical heat was the norm for southern England.

Meanwhile, many of the early flowering plants themselves evolved into broadleaved, flowering trees: the oak, the hazel, and the beech; the magnolia, the baobab, and the eucalyptus. Some became as big as conifers; others tucked in under the canopy and used different techniques to thrive. Using long vessel elements for transport, they rely on fibers for support and axial parenchyma (sometimes known as xylem parenchyma) for storage and embolism repair. A greater number of xylem parenchyma cells are found in tropical species, and they are particularly abundant in the most plastic of all plants: the lianas and stem succulents.* A beautiful illustration of this is *Machaerium floribundum*, which I saw growing in Brazil. While the tree form of *Machaerium floribundum* can grow up to 82 feet in height, with yellow pea-like flowers hanging from delicately drooping branches, I saw it growing as a liana, twirling up a larger tree, and in this form it has approximately twice as much axial and radial parenchyma in its xylem.† This parenchyma allows the plant to be flexible and mobile—in comparison with the stiff tracheids of the conifers it's as though it has been liberated from wearing a whalebone corset—and to thrive on very little light and therefore photosynthesis.

When annihilation, in the shape of the Chicxulub asteroid impact, hit the earth, and photosynthesis was impossible for two years, it was the sprouting, earthy angiosperms, trees that were able to resprout from their meristems, and not the towering gymnosperms like the sequoias, which could only regrow from their burls (special nodes for recovery), that developed to create the most diverse landscape on earth. An open forest of gymnosperms turned in a very short window into a multilayered forest of angiosperms, and they now sit as the basis of the evolution of terrestrial biodiversity.[8]

A young and brilliant professor of botany at the University of Michigan, Monica Carvalho, was the first person to recognize that modern forests were produced by the extraordinarily dramatic impact of the Chicxulub asteroid.[9] The great diversity of the rainforests arose out

---

* In addition, plants representing different growth forms show varied proportions of xylem parenchyma.
† Remarkably, the amount of parenchyma tissue fraction in stem succulents can be as high as 70 percent.

of the two years when no photosynthesis could happen on earth. A great cloud of oil and tar burned, and the smoke blocked out the sun. It took 6 million years for diversity on earth to recover, but recover it did. By using autochthonous assemblages of fossils (those that sat within their ecological webs and could be seen as part of an ecosystem), a snapshot rather than a great sweep across thousands of years and hundreds of miles, Carvalho and her team could look at the fossil leaves and how they developed. They could estimate the rainfall by looking at leaf imprints like drip tips, and the amount of leaf mass made, and, by looking at the size of the stems and branches that were locked in place, get an idea of how much carbon dioxide was being made into leaf and biomass at any given time—how productive each square of forest was.

The change was profound. Imagine a forest with monkey-puzzles growing across the landscape. All is deep-emerald green; the trees are old and wide at the base, with bark like elephant skin, the trunk bare for the first 7 feet, and resin oozing out of the nubs of the discarded branches. Higher up, the languorous sweeps of the branches make a cone shape, and each branch is spiraled all round by tightly packed scales like an emerald pangolin's hide.

This is the theme, on which there are variations, a fugue of shapes marching across the landscape. Some of the other conifers in the forest will have smaller branches, with the spirals around the trunk more pronounced, rather like the beautiful Norfolk Island pine, *Araucaria heterophylla*. Some will be larger, with flakier bark more like cypresses or spongy bark like the giant redwoods. These are the keystone species, and underneath are ferns and horsetails and cycads and fungi, thick on the ground and piled up into leaf litter. The ferns are much like those we know today: heavy with spores and branching symmetrically around a central point; the cycads sit on their bulbous stems and also carve out a circular swathe of sun with their stiffer fronds. Swarms of insects occasionally settle on an area and almost completely defoliate certain species in intense cycles of boom and bust. Huge dinosaurs like the diplodocus occasionally wander through and from their vast height uproot and consume a patch of trees, trampling and munching. The trees, adapted to this, will regrow from burls or shoot up into the space that is left, and so a stable but dynamic ecosystem holds firm.

Then suddenly the asteroid struck, and all was darkness and confusion, mushrooms and rot. The green shade of the trees became a gray gloom as the dust cloud generated by the impact blotted out light from the sun for two years, and the saprotrophs (literally "eaters of putrid things") boomed. The area of South America that is now the Amazon rainforest became a death zone, with rotting wood and dead trees, starving dinosaurs and dead insects. The climate remained wet and moist, but after two years of oppressive darkness a huge number of plant species had become extinct. The monkey-puzzles in particular had suffered hard, and the cycads too had struggled, just like the ferns, as ash and dust from the impact rained from the sky and spread a gray layer across the earth and as the dead trees dropped leaf litter and branches and finally fell.

This blanket of ash and death, which forms a distinct layer in the geographical record, smothered the ferns and cycads, but the flowering plants grew through it. Up came the star anise plants and their relations, the *Amborella* struggled on and spread. *Chloranthus* clambered over what was left of the monkey-puzzles, and a plethora of other flowering plants came in to evolve into different niches with larger leaves and flexible branches. The stately galliards of the araucarias, well spaced and single canopied, gave way to a mass tango, angular and fluid, in which a large family of plants, the legumes, fixed nitrogen for everyone and many layers of canopy could intertwine. Over the next 3 million years the rainforest that we know started to develop.

When we think of the productivity of the rainforest, we can see it as a messy sort of democracy, unlike the feudal systems of the past, in which strict hierarchies were maintained. The revolution that led to this tight-knit interconnectivity is obvious: two years of ash, soot, and mushrooms. But given the similar climate before and after, why didn't new conifers move into the shoes of the old?

One possible answer is the death of the dinosaurs and other large herbivores. Once they could have maintained constant disturbance patterns but, after being wiped out by the asteroid, they were no longer there to consume huge quantities of biomass. Consequently a race for light began. There are parallels here with the use of large herbivores such as bison and cattle today to maintain a patchwork in a rewilding context.

A second answer is that stable forests like the coniferous forests, which had dominated for millions of years, very quickly weather the rock underneath them, leading to weathered and extremely infertile soils. Conifers are better adapted to infertile soils, but the ash that rained from the sky after the asteroid impact acted like an incredible boost of high-phosphorous fertilizer, allowing angiosperms to swamp conifers. Those angiosperms like the legumes that could fix nitrogen had nothing holding them back, and it was around this time that the Fabaceae, now everywhere in the rainforest, became an important family. The modern equivalent is the dust of the Sahara, enriching with phosphate the poor soils of the Amazon rainforest.

The third and final probable reason for the change was the selective extinction of many conifers and other gymnosperms. The auracarias were not diverse; they were keystone species that structured and shaped everything underneath them. Inflexible and slow to change, they were much more likely to become extinct than the angiosperms, which could twist and turn out of trouble, picking up the last and first rays of light from a clouded sky. Importantly also, the angiosperms were more acrobatic with their genomes, able to splice, selectively express, and duplicate their DNA at will. This allowed them to be "shaped" more than conifers and is why even very long-lived flowering plants, like the oak tree, can adapt fast and flexibly to new environments.

On the coast of Namibia and in southern Angola a gymnosperm survives in unusual circumstances, changing little over the millennia. *Welwitschia mirabilis* is a slow-growing, cone-shaped knob of wood embedded in desert sand. From it grow two ragged leathery strap-like leaves, which survive both searing heat and wet mists off the cold sea and can continue growing for a thousand years over the whole life of the plant. Male and single-seeded female cones are formed on branched stems that emerge from between the leaves. *Welwitschia* was formerly more widespread, as its pollen has been found in North America, Portugal, and Brazil.

All this is to say that, having developed under the strict repression of tree canopy, the angiosperms, this new set of plants, were easily able to scavenge what was left of life after the asteroid. Many of them became trees themselves, never quite as tall or ancient or structurally shaped as the gymnosperms, but enormous canopy trees nonetheless. They also became alpines,[10] growing on the tops of mountains, water plants, and plants of the open veld.[11]

The better use of light meant that layers and layers of vegetation could grow up, biomass per acre boomed, the canopy became tight and closed, and the rainforest we see today began. There are illustrations of this asteroidal change in the beautiful landscapes that survive in Brazil. By the sea in the cooler south of Brazil, an *Araucaria, Araucaria angustifolia*, also known as the Brazilian auracaria, still stands in forests that are technically known as ombrophilous, or shade loving. It stands tall in an open-canopied forest, controlling the plants all around it, the firm keystone species in a sea of mist-loving angiosperms, laurels, myrtles, and many plants that were characteristic of the circum-Antarctic flora. This forest, and forests like it, are as close as we can get to the land where dinosaurs roamed, with single-stemmed trees branching in fractal patterns, their branches twisting interrogatively out into the air in tails of spikes.

The forest is also a hodgepodge of what came afterward. Grasses have replaced ferns, and the only dinosaurs that remain are birds, like the azure jay (*Cyanocorax caeruleus*), which eats and distributes the 2-inch-long nut-like seeds released when conifers' cones are mature. There are few epiphytes, the ecosystem is poised, inflexible, and the trees shape the plants around them abiotically, through soil and water, rather than with the physical impact of plant against plant.

At the other end of the scale is the Amazon rainforest, and in particular the area around the Rio Curicuriari toward the Colombian border. The Rio Curicuriari is a tributary of the Rio Negro, the black stream of poor, humic acid-laden water that joins the Amazon at Manaus. The soil is ancient and weathered, almost pure silicate sand, and yet there is an incredible level of diversity, layers upon layers, and some very large trees, such as the very rare duraka, *Aguiaria excelsa*, which grows up almost 200 feet out of the thin soil. Most of the diversity is driven by legumes, those podded smaller trees with their feathery leaves and mercurial ability to reach into sunlit spaces and fix nitrogen into even the poorest soil. Undergrowth is sparse, as little light ever reaches the forest floor, so young trees play the role of undergrowth, growing slowly and biding their time for years, until a tree falls down and the light floods in; then they race for the surface. The forest floor is scattered with pods, big and small and with ruffled edges, the sign of nitrogen fixers, and the trees flower in a million different ways, intertwining with one another and other plants to spread seeds, develop, and grow.

Plants that are not trees tend to be epiphytes, growing on branches high in the tree canopy: orchids at head-height with deceptive, pendulous white petals and scurfy tubular stalks; climbing and creeping lianas with heart-shaped waxy green leaves; stiff, pineapple-like bromeliads and fresh-green philodendrons; hairy, tubular-flowered gesneriads and leafy, clinging cacti, climbing trees to find the light and fitting round them in ever more innovative ways.

Giovanna, a botanist who was my host among the Baré tribe, pointed out some of these to me. One of them we saw on our first morning of going into the rainforest together to look for the showy pau-brasil, the brazilwood tree (*Paubrasilia echinata*), with feathery leaves and bright red and yellow flowers. We had moored the canoe in which we had gone upriver in an *igarapé* (a shallow channel running off the main river into the forest) and scrambling up the bank we saw a string of red, flattened leaves clinging round the trunk of a small pau d'arco tree (*Tabebuia impetiginosa*). To me, the clinging, transfer-like leaves had an unpleasant leech-like look, but Giovanna was thrilled. "*Isso*," she said, "*é Strophocactus, flor de Margaret Mee!*" It was the Amazon moonflower, the rare *Strophocactus wittii*, one of the trickiest of all plants to find in flower. It does not bloom every year, and when it does, only for a single night at the end of April or beginning of May.

Half closing my eyes, I knew what she meant about Margaret Mee. In my mind's eye I could see a picture in a gallery in the Royal Botanic Gardens, Kew, in London, 6,800 miles away, where I used to be taken as a child to stay out of the way while my father was working in the herbarium. This was a painting that showed the flat leaves uncurling away from the trunk, stalks springing up and holding out huge white and pink flowers to uncurl in crepuscular darkness to the moon. To paint the picture, the botanical artist Margaret Mee had to plan a two-month trip from Rio de Janeiro, where she lived. It was her fifteenth up the Amazon, and she was still recovering from a double hip replacement. She had to search the flooded forests of the Anavilhanas Archipelago for a fortnight before she found a cactus in bud. She waited until nightfall for it to flower and, as the moon rose, suddenly it opened. "An extraordinary, sweet perfume wafted from the flower," she wrote, "and we were all transfixed by the beauty of the delicate and unexpectedly large bloom, fully open in an hour." Mee thought the flower would be pollinated by a large moth or a bat, but nothing came apart from small flies. "Our intrusion had deterred the pollinator," she lamented, "upsetting the delicate balance between the flower and its pollinator that had taken tens of millions of years to evolve."[12] It was her last expedition, and she died six months later.

In many cases conifers have shaped flowering trees, which have themselves shaped flowering plants around, on, and underneath them. It's easy for trees to do this because they have absorbed the lion's share of carbon and are so structured, so big, that they become a three-dimensional habitat. But they also often shape plants more directly and deliberately by exuding chemicals. In the oldest, slowest rainforest in the world this double shaping takes on an otherworldly beauty.

The Chilean rainforest is not tropical but temperate. The poet Pablo Neruda called it "the vertical world": an entire new dimension of the planet, living, breathing, and interconnecting. The largest and longest-lived trees in Neruda's Chilean rainforest (and incidentally in the southern hemisphere) are *Fitzroya cupressoides* or alerce trees. Walking among these huge conifers was like walking through a pillared temple of immense antiquity, on a floor carpeted with brown discarded scales, tiny, scattered cones, and small twigs. Massive trunks (Darwin recorded the circumference of one as 41 feet) are covered with thick, silver, flaking bark designed to repel fire; rather than the gravity-wrinkled elephant skin effect of the monkey-puzzles, the oldest have a knobbly effect like a lathe-turned bedpost. The leaves are scales, making canopies that seem somehow inadequate for the vast trunks, reaching far up away into an often gray and raining sky. Alerces connect heaven and earth, pulling phosphate, nitrogen, water, and other minerals 200 feet up to capture carbon in the sunlight and make sugar, before whooshing it 200 feet down again to store it and power growth. Alerces are well proven to live to 3,613 years old, but the oldest is suspected to be Gran Abuelo, or great grandfather. Possibly 5,500 years old, it is described as a waterfall of green.

Beneath these megaliths even the largest of the broadleaved trees look unimportant. Magellan's beech, *Nothofagus betuloides*, can grow to 100 feet, but beside the alerce it fades into insignificance. *Nothofagus* has small oval leaves, deeply veined and stiff, and smooth brown bark that twists slightly as the tree grows up. Directly underneath it grow laurels, *Drimys* with large almond-shaped emerald-leaves and white flowers, myrtles with fragrant, flaky white-and-orange bark, multibranched and shrubby, and the bee- and moth-pollinated *Eucryphia*. Under the laurels grows the Magellan fuchsia, *Fuchsia magellanica*, with red and purple flowers, and little else—moss, ferns, and the occasional reed.

Most plants, however, are driven by lack of light from the ground up into the trees. A study conducted on six *Fitzroyas* found that fifty of eighty-three plant species associated with the trees were living in the canopy: lianas, filmy ferns, and two other species of tree. For Neruda, trees were the arteries of the forest, and it didn't take long for plants and animals to tap into these vertical rivers. Some of the adaptations to a perfect vertical city are showy. Others are more subtle and interconnected. *Lapageria rosea*, the exquisite waxy-pink bell-flower that is the national plant of Chile, uses its smooth stiff stems to wind up trees until it can display its flowers in the sun, essential to attract hummingbirds for pollination. The same hummingbirds are dependent on the presence of a mistletoe, the tree parasite *Tristerix corymbosus,* to survive throughout the year, as it is the only plant with the pillar box-red tubular flowers needed by the hummingbirds that can afford (with a little help from the trees around it) to flower in the dead of winter. As a result, trees such as *Nothofagus* produce chemicals that actively encourage the establishment of this parasite, supporting the mistletoe and through it the growth of *Lapageria,* which will then be useful to the tree, protecting the trunk from damage and ensuring a balance in the canopy.[13]

Alexander von Humboldt, the nineteenth-century German traveler and botanist who spent much of his life exploring South America, believed that the massive diversity of flowering plants in the rainforest indicated a highly stable ecosystem. He was completely wrong: it is in fact one of the most unstable ecosystems in the world. The flowering plants, which now make up 90 percent of all known plant species, are able to capitalize on this instability because of their less regimented structure. Evolving a hundred million years ago, before the obliterating catastrophe of the Chicxulub asteroid, the dark, wet, and disturbed environment under the conifer canopy made them intrinsically flexible, adaptable, and creatively chaotic. Filling every space, they fixed carbon and produced biomass like never before, and have evolved to live in almost every habitable niche across the world.

Even those plants that appear to have nothing to do with trees have been shaped by them. At 12,800 feet at the top of the Tosor Pass in the Tien Shan Mountains, normally hidden under the snows that divide China from Kyrgyzstan, lives a plant with flowers that look like a slice of the moon. It is the snow-globeflower, *Trollius lilacinus,* and although it grows across Siberia and Mongolia the nearest tree—in this case the Tien Shan spruce, *Picea schrenkiana*—is miles away, sheltering more than 3,000 feet lower in the bottom of a gorge. For nine months of the year the globeflower huddles under the snow on the mountaintop, occasionally falling with landslides off the side of the mountain or tilted by an earth tremor to leave it facing sideways. Even before the snow has entirely melted it puts up one or two tentative stalks, then three, and then five globes, which burst into a mass of multipetaled ice-blue flowers. A little later it will put up leaves and, photosynthesizing fast, ignoring the mass of perils surrounding it, suck the icy meltwater up into its vessels, providing a warm spot for a pollinating fly to rest. In the blinding white of the snow, surrounded by mountain peaks rising out of the cloud, it seems impossible for the globeflower ever to have seen a tree, let alone be shaped by one. Deep inside its stems and leaves, however, the embodied memory of that evolutionary time remains. The flexibility that allows the stalks to ease the globeflower's buds past shifting shards of stone that come down with the meltwater, the vessels that allow it to hunker down under the snow and then sprout up, like lilac smoke, into the warmth of the sun, and most of all the flowers themselves, which will transform into a head of small seeds rapacious for bare ground—all carry the imprint of the forest floor.

CHAPTER 7

Trees shaping animals

Trees shape animals chemically and physically, and mainly they use them to move

And all our yesterdays have lighted fools
The way to dusty death. Out, out, brief candle!

William Shakespeare, *Macbeth*

The hair of a single three-toed sloth can contain more than 900 bee-tles. It can also carry entire populations of highly specialized moths, along with unique algae species that grow on the hair and tinge the sloth a delicate shade of Kendal green.[1] As well as acting as camou-flage, the algae provide the sloth, and the moths that live in it, with lipid-rich food to augment their diet of tree leaves. Moths provide the algae and the sloth with a source of nitrogen-containing protein, and that nitrogen is solemnly recycled by the sloth once a week as it descends to the ground (slowly) and unhurriedly defecates in a hole that it will carefully scrape and re-cover at the base of the tree. This apparently unnecessary and dangerous procedure allows the moths that the sloth carries to complete their life cycle in the ground. The advantage to the tree is fertilization: the great flood of nitro-gen and phosphate, delivered accurately onto its roots, that the sloth provides.

So how do trees shape this peculiar interaction? The answer is through chemicals, in particular the chemicals involved in the sloth's brain. Individual sloths will often commit to a particular tree as a home base and key leaf producer, and that host tree will be one of ninety species. It might be a cacao or a *Dinizia* tree, but often it's the

Cecropia or trumpet tree, a common tree in the rainforests of Central and South America, and one that displays an extraordinary number of adaptations useful to animals. Cecropias grow up tall out of disturbed land, producing huge fans of multipinnate leaves that flop out crumpled and silver and gradually expand at the top of their white trunks like sweet green umbrellas. These juicy leaves have glycogen structures at the base of the leaves that are designed to attract ants. And the ants, while protecting the tree from insects and vines, do not harm sloths. When cecropias fall you can see the reason that they can grow so fast—their trunks are hollow and compartmentalized, thrown up hastily like a scaffold. The need to live in trees like *Cecropia* and others with not particularly strong branches means sloths must not be too big. Their energetic constraints, therefore, are huge, and they must move incredibly slowly, expending just 460 kilojoules a day—the equivalent of 110 calories or a small baked potato. Descending to fertilize is an enormous expense of energy.

Sloths are one of the few larger animals that manage to live on nutrient-poor leaves alone, so they have been molded more comprehensively than most by their tree lifestyle, to the extent of having become like trees themselves. If there is a strong animal contender for Tolkien's Ents it is the green, gentle sloth, which can live more than thirty years and whose trip down the tree and back is like a human having to run 5 miles just to go to the bathroom.* Whereas sloths have almost become tree, most animals are shaped by trees to become even more animal and do everything that is non-tree.

In particular, trees use animals to move. As primary producers, trees have shaped the size and shape, as well as the senses of sight, smell, taste, and sometimes touch, of the animals they interact with. The Brazil nut tree, for example, has gradually inflated the strength of bees in its area of the forest, supporting only the huge black euglossine bees that can force their way inside its flowers. As a result

---

* Sloths diverged from other mammals about 100 million years ago along with the anteaters and the armadillos, but only evolved into something approaching their current form 60 million years ago. The further a sloth adapts to its tree habit, the slower and more green it becomes.

of their size the bees can travel long distances between the huge and widely spaced trees. The *Ceiba pentandra* tree has shaped some bat's noses so they can only be used to access nectar. Blocking the bats from other sources of food forces them to travel miles between trees in order to find one flowering that day—in the process transporting pollen and optimizing genetic flow. The pau-rosa tree, *Aniba rosodora*, from which humans get one of the ingredients for Chanel No. 5, has persuaded toucans to become completely peripatetic to move tree seeds from place to place. Cacao trees of many species invest a huge amount of energy in spiking their fruits with caffeine and fats and sugars to allow mammals to travel further and harder in pursuit of more fruit. In general a larger animal is cleverer and requires a larger and more tangible reward, but they can nonetheless be chemically and physically shaped by that reward. Larger animals also have the advantage of being able to move further and are therefore better suited to spreading seeds. Insects, on the other hand, are easier to manipulate, because their brains are simpler, so they're more often used as a conduit, spreading pollen for miles, like the personal courier of DNA for a specific tree.

Imagine you are a tree trying to attract a bee. In the general scheme of things this wouldn't be a problem: bees would forage through the woods or over the fields that surround you, and you would be "searched" as a matter of course. But now imagine that the pollen you need *just isn't arriving*. The once-thriving population of trees you were part of has been killed off by climate change, and you have survived only by sheltering in a steep gorge.

This was what happened to the horse chestnut tree about 3 million years ago when cooling began at the start of the Pleistocene. Members of the species survived in Bulgaria and in the Vikos Gorge in Greece, sheltered deep behind overhanging stone, and they reproduced by tricking bees. The gorge acted as a refuge, but only the promise of a very large reward would tempt bees down to its floor. In order to attract bees during the critical three-week flowering period, horse chestnuts evolved a startling visual effect, using their huge white flower spikes. Rather than allowing oxidation of polyphenols within the petals, which would cause the flowers to go brown and fall off once pollinated, chestnuts maintain even their pollinated flowers in pristine whiteness, smothering the tree in white and making it irresistible to bees. However, to make sure of attracting bees only to the unpollinated flowers, they have evolved to slightly alter the tiny splash of color at the center of the flower, called a nectar guide, once pollinated. By controlling the nectar guide so it changes from a strong yellow signal to a weak crimson one, they indicate an absence of nectar, and thereby control the bees' movements.

A couple of years ago I went to try and find these chestnuts. I'd wanted to see them in flower for years, ever since I'd heard the story of their survival and spread, and year after year the repeating alert in my iCal, "Horse chestnuts in flower in Vikos," had mocked me. Finally, everything was aligned. My friend Josh, not an easy man to pin down, who lives in Ano Pedina, just a short walk over the mountains, was at home, I was free, and the chestnuts were presumably flowering. This area of Greece is a limestone landscape. The village, with its limestone houses staggering under the loads of their stone roofs, sits on the side of a hill next to the plain of Soudena and is famous for its nightingales. We listened to their voluptuous arias every night. You could hear them in groups, somewhere out in the darkness, sometimes nearer, sometimes further away.

It was drizzling when we set off for the gorge, humid and overcast, but the greens of spring couldn't have been better. There were a million beautiful shades, with the occasional adumbration of a trunk, different trees at different stages. Wild pears offered occasional impressions of white, while the firs stood dark and black against the mountains. In the distance the snow-streaked peaks were white against a gray sky, and across the mountains occasional clouds hid the green. The clouds were thickest toward the village of Monodendri (in Greek, literally "the place of a single tree") at the head of the gorge, and the steep limestone cliffs showed the path of the gorge through the blanket of green.

Although the gorge is just over the hill, there are few places where you can access it, so we drove to the Kokkoris bridge, an eighteenth-century stone semicircle spanning a spur of the gorge. From there, as if through a triumphal arch, we walked into the gorge proper. Instantly the pale green of horse chestnuts started to appear. Not in profusion: just here and there along the gorge, mixed in with the trees that make up the scrub and woodland along the river. They grew out of cracks in the cliff edge and then rose up vertically, thick candles of flowers all at different stages of transformation, from green bud into white splendor.

Leaving the car and walking under the bridge was like stepping inside a Byzantine basilica: dark next to the steep walls and occasional pillars of stone, rays of light shining from above through the trees, the light candles of the horse chestnuts leading to the iconostasis of the bend where the river disappeared. Even in the cold northern Greek spring you could feel the humidity and warmth of the gorge, a microclimate where a tree could hunker down to survive.

The gorge cuts through a cross-section of the mountain, and from it you can see other mountains, still snow-capped and teeming with botanical treasures. They have euphonious names and a distinct character: Tymphe, the one with striped snow; Smolikas, the great hulking mountain; Mitsikeli, the smaller one with many peaks to the south. The gorge running through Tymphe looks as though it must have been formed by some sudden ripping apart of the earth, but in fact it was the gentle action of a river over thousands of years, melting through layers of limestone accreting across millions of years. Eocene limestone formed under the sea makes up the top layer, and the river runs through the Cretaceous to the deepest layers of gray Jurassic dolomite. The shelter is absolute. The gorge twists and turns, and every smell is intensified, humidified. This is where a tree could hang on in an age of ice and snow, even as the last of the laurel forest that it grew with was dying out.

Here and there in the gullies furrowing down into the gorge horse chestnuts hung on through the Ice Age. But the gorge must have started forming at some point around then, which means it would have been a race against time for the trees to survive before they started to have a true microclimate to hide in. Their survival relies, hugely, on the right pollinators surviving. Bees were the pollinator of choice, and horse chestnuts have honed a way to shape their behavior so obvious that even humans find it hard to miss.

At the head of the gorge the rain clouds were thick and black. We walked on up the gorge, noting every horse chestnut, as well as the hornbeams, the woodpeckers, the occasional nightingale, the orchids. Most of the horse chestnuts were still in bud, not fully out, but finally, crossing another of those perfect gray rainbow bridges, we saw a beacon of white: a horse chestnut in perfect, radiant flower against the cliff. We crossed the precipitous bridge and jumped from stone to stone across the stream to get right up under its branches. There it was: the newer flowers with their flash of yellow, the old flowers with their splash of red, and a bee moving from candle to candle, yellow-hearted flower to yellow-hearted flower, just as it should be.

As soon as a newer, yellow-hearted flower is pollinated, a chemical reaction takes place that converts the yellow into red, but leaves the white petals of the flower relatively unfaded. We don't know when this reaction evolved, but the result is to magnify the pollination signal from far away so a bee sees pure white and will carry pollen all the way up and down this tortuous gorge route, allowing for a movement of genetic information that helps the small population of trees to thrive.* Further up was another, older tree, not quite in flower, and higher up the gorge, on the rock underneath its shade, was a purple covering of the Serbian phoenix flower, *Ramonda serbica*, a strangely primitive-looking plant with deep emerald leaves and perfect five-petaled flowers, one of the prettiest survivals from the Tertiary. *Ramonda* is pollinated by bees and offers only pollen in return.

Horse chestnuts provide bees with nectar and pollen, but they also trick bees into designating them an even better foraging source than they are. Where humans see the white candles of horse chestnuts

---

* There are examples where inbreeding can be used to direct evolution, but this is normally a risky strategy—outbreeding is almost always the most effective way for trees to evolve and stay genetically strong.

standing out against the light green, with closer, subtle specks of yellow or red, bees see an additional dimension. The large white signal of the flowers on the valley side turns, on closer inspection, into a dazzling strobe show, with patterns of UV lights reflected from the stamens and anthers glowing deep like a pile of gold into the middle of the yellow-centered flowers. The red-centered flowers, on the other hand, will have just a small strobe on the anthers—the center of the flower—and to a bee the nectaries will be a dead black.

Bees have been modified by trees to see no red: only UV, blue, and green.[2] They can see yellow and orange but, like people with aphakia, such as Claude Monet in his blue period post cataract surgery, the purple, violet, and blue that makes up most of their vision is irradiated by an extraordinary light of iridescence, a mosaic of glowing spots of intense gold that shine beyond the large patches of purple, into the glorious patterns of a million flowers. Even under overcast skies bees live, like the Byzantines, in a world of glowing colors and shafts of light from beyond the world of violet rays. Their large compound eyes can see light polarized by air molecules in the atmosphere and can recognize the patterns of polarized light in the sky, allowing this to be the framework for the foraging map they will relay through their wagging dance. Three small ocelli on the top of a bee's head will transmit information about UV light and sunlight. Studies of what happened when bees were deprived of UV light found they stayed in the hive until only severe food shortages and starvation forced them out. Life without ultraviolet was too dull for bees to go out. Without tree art, food was almost too boring to eat.

The strobe show of a flowering horse chestnut is created by esculin, a fluorescent molecule, and the same molecule, when used as a poison, is responsible for another dimension of horse chestnut survival in which horse chestnuts manipulate animals to move on their behalf. The seeds of the horse chestnut, conkers, ripen on the tree in a spiky capsule that pops open in time, releasing the shiny brown nuts. These nuts will sometimes fall into the river that runs through the gorge and find a new home in a periodic flood, protected by their hard skin and smooth roundness, but more often wild boar and deer and squirrels will eat the conkers or bury them in a cache, moving them for considerable distances. The seeds contain poisons, essentially soap—alkaloid saponins—that will rip apart the fats inside the animal, and glucosides like esculin that release the coumarin poison if not properly processed or eaten in too great a quantity.

In the rare cases that human cultures eat conkers, they bury the seeds in damp soil, which encourages the poisons to leach into the ground. In addition, the nearer that a fruit is to germinating the less poisonous they become. Animals use just this mechanism to avoid being poisoned, but it's a beautiful example of a timed chemical mechanism: the poison ensures that animals will bury the conkers and then leave at least some of them to germinate. Just like the huge and flashy pollination signal, this mechanism for using animals was probably essential for the horse chestnuts' survival against the odds.[3]

It's a pretty solution to the problem of attracting bees, and a similar mechanism has evolved independently to allow trees to spread their pollen over steep gorges in China, where during pollination season the handkerchief tree (*Davidia involucrata*) can be seen as a white flash among the valleys of Sichuan. The tree has developed enormous sepals that wave white in the slightest breeze, and by doing this it has harnessed the movement of bees, which travel miles across the mountains and through steep gorges to find it. This and other such techniques have locked in bees and other insect pollinators to rely on trees. Most need pollen and nectar from trees' flowers to survive, particularly at times in the year when other flowers are absent.

Shaping bees so that they see wavelengths longer than purple was fundamental in the development of pollination and flower shapes, but trees have also shaped the sight of birds and bats. Many bats see reflected UV at very high intensity, essential for those who feed on nectar. Most strikingly, bats are used by banana trees to spread seeds, and the tree controls the bats not only by making the bananas sweeten only when they are ripe, but also by patterning them with brown spots on a yellow background like a leopard. This camouflage under the dappled light of other trees may protect the fruit from animals like monkeys, which will greedily consume the fruit and not spread the seed, but to bats the spotting has the opposite effect. Under UV light the brown spots (caused by chlorophyll breaking down) fluoresce, and the banana strobes like a cartoon cow. The banana tree has in effect given its offspring wings.

Bats find ultraviolet vision indispensable in the twilight conditions they tend to operate in, but like most mammals that evolved to see well in the dark, they are not particularly good at distinguishing between colors. Birds that live in forest environments also often develop the ability to see UV light: it allows them to navigate flight patterns through dense rainforest canopies better. The upper surface of a tree's leaf is often covered with a waxy cuticle to protect it from rain. This wax will reflect UV light, appearing brighter to a flying bird than the underside of the leaf. Many trees harness this adaptation by covering their berries with wax, not only protecting the seed and fruit but also making them highly visible to the birds that would eat the berries and disperse the seeds.

It was dawn in Guatemala. After a month digging and scraping and washing pottery and floating tiny fragments of plants in Belize, we'd crossed the border to visit one of the great Mayan sites, Tikal, and had forced ourselves out of bed in the dark to climb up Temple IV and see the sunrise over the forest. As we walked through the fabulously sticky forest a big gray bird was hopping from tree to tree in the first light. On top of the temple the mist cleared and the pyramids appeared, impossibly huge and beautiful across the landscape, and the astounding wealth of flowering trees, ceiba and cedar and mahogany, in the forest below us gave everything a light shimmer of pale yellow and white. Parrots and other birds flew across the canopy, and the tousled, messy look of the trees gave an impression of irrepressible life.

On the way back I saw the same bird, and—no longer gray in the crepuscular half-light—this time it was unmistakable. The huge beak, the beady eye, the vibrant coloring, a toucan, unmistakable from the Guinness advert but much more magisterial. In Brazil toucans eat and spread the purple drupes of pau-rosa, the Chanel No. 5 tree, as well as the glossy deep-black fruits of the jaboticaba and those of the açai palm, also deep purple and waxy. They're attracted to them because deep-purple fruits reflect UV light, appearing to the toucan as gold-covered sugared almonds would to us. Toucans also love the orange-pink sweetness of papaya and the pink of the strawberry guava (*Psidium cattleyanum*), as well as the bright-red figs of *Ficus calyptoceras*.

Relationships between birds and trees began at least 50 million years ago, and may go back as far as 170 million years, when dinosaurs, the forerunners of the birds, glided between the trees with some rudimentary and innovative feathers on their backs. Even before that, the cycads were producing brightly colored fleshy fruit surrounding highly toxic seeds that appear to have spread all over the globe by hitching a lift on dinosaurs' backs.[4] Cycads can be trees or shorter plants and in one extraordinary case an epiphyte, the only gymnosperm known to grow on another tree. They have stiff spiky foliage in a fan shape and developed serious toxins to prevent this from being eaten by dinosaurs. The neurotoxin cycasin, a glycoside similar to esculin, and the carcinogen macrozamin will together prove fatal to most animals, including humans, that consume them. In cows the disease is called the Zamia staggers, in most cases causing instant vomiting and quick paralysis and death, while in the longer term, even at low doses, cancer and neurodegeneration will make sure the animal foolish enough to graze on the leaves will never reproduce. Only a few insects can survive, one of which, a Lepidoptera larva, weaponizes cycasin against its own predators.

Cycads are an interesting model, because we can look back in time and observe both their rise and fall. Dinosaurs were formidable herbivores, who would have favored the cycads over gymnosperms like the gingko or *Araucaria*, which didn't have poisons as lethal as the cycads, by essentially mowing over them like a conservation-grazing animal. Using dinosaurs to move their seeds around enabled cycads not only to spread geographically but also to gain resilience. They could cross-pollinate with different gene pools, move to different niches in response to shifts in weather patterns or climate, and avoid insects and microbial pests that might otherwise migrate onto them. The stones in a sauropod's gut, the gastroliths they used instead of chewing their food, acted as polishing stones for the seed, cleaning them of pathogens and lightly scarifying the cases so they could germinate with ease. In fact, it may have been because of their diet of cycads that the sauropods had sharp, nipping teeth and microbial digestion, rather than the later grinding molars of the hadrosaurs, which would have crushed the seeds and poisoned themselves in the process. Most importantly, the system the cycads developed of a large and poisonous seed surrounded by sweet and succulent flesh ensured that only dinosaurs large enough to go the distance required by the cycad could survive swallowing their fruit. The glycoside the cycad developed acted as the iron fist inside the fruit's velvet glove. There is a school of thought, however, that says the cycads made use of dinosaurs only too well. At their pinnacle they were the most abundant gymnosperm on earth, but the symbiosis also left them unable to compete with the angiosperms once the dinosaurs had been wiped out.

Back in Belize I went to try to see a surviving but very rare species of cycad only reported in 2009. It was called the reclining cycad, *Zamia decumbens*, because it lounges gracefully over the soil on a stem a bit like a looping caterpillar and is found only in one or two sinkholes and the so-called Maya Mountains, a limestone range that runs toward and along the Guatemalan border. The mountains are catnip for botanists. Bruce Holst at Marie Selby Botanic Gardens in Sarasota, Florida, tells of a conservation trip in 2007 between two Category 5 hurricanes, when the team had to be flown out by helicopter from the most inaccessible part of the forest, but he still saw the greatest diversity of plants he had ever seen in his life. The area's glamour is enhanced by the fact that it is a demilitarized zone; the controversial Belize-Guatemala border is stabilized by the presence of British troops on one side and US troops on the other. After asking for the locations (which are otherwise kept secret because of the extreme rarity of the cycad), we opted for one of the sinkholes and, after a considerable amount of planning, off we trekked.

We never got there. There were no paths, and the dense forest completely obscured the route we were trying to follow. We should have come extremely close to the top of the sinkhole, but despite one of our party having visited before it was impossible to find. Every-thing had changed. In some ways I was relieved. I saw it in cultivation in a botanical garden: a stem flopping along the ground, red fruits held up on a stem from which they could be neatly nibbled. To have seen it in its sinkhole would have been like viewing the last remnants of a dead empire.

As the rainforest that grows up around them has developed and spread the cycads' distribution has waned massively. Without dino-saurs to spread their seeds they have become inbred and their popula-tions fragmented. Other animals have stepped in to help. In the case of the Kananga cycad (*Encephalartos poggei*) in Angola, the new animal that helps its distribution is the only remaining representative of the megafauna, the African elephant, which swallows the large mass of seeds without chewing it, passing the poisonous seeds a couple of days later. The seeds of the rare cycad *Dioon spinulosum*, which grows up to 40 feet high over steep hillsides in Oaxaca in Mexico, are rarely and badly dispersed by the Mexican black bear. Mostly, however, large birds disperse the seeds. The Australian cycad *Macrozamia riedlei* has seeds that are eaten and spread by the moyadong, a species of par-rot, and the emu.

If you live in Western Europe or Scandinavia it can be easy to for-get about tree flowers. In Belize or Brazil or the Philippines, Pakistan or Madagascar, you cannot forget. In the Bagh-al-Jinnah in central Lahore, the magnificent red flowers of the bombax (*Bombax ceiba*, not to be confused with *Ceiba pentandra*) create canopies of radiant color. In the Philippines the mangkono tree, *Xanthostemon verdugonianus*, puts forth its red brush-flowers, and in the streets of São Gabriel da Cachoeira, Brazil, the pavement is strewn with flashes of hot pink so intense it looks like a Holi celebration, from the anthers falling from the street trees transplanted from the other side of the globe, where it is native in Java and Sumatra as well as Australia. *Syzygium malaccense*, the Malay apple, as it is sometimes called, is also one of the canoe plants taken across remote Oceania as a crop and was intro-duced to Hawaii 1,700 years ago. It doesn't only have ravishing pink flowers but also large red fruits, almost too large for birds to spread. In Madagascar the baobabs produce orange-yellow trumpet-shaped flowers of great beauty, and in Angola the trees of the mpingo often have ravishing flowers.

Insects can't see red, so we can be fairly sure that all these trees developed to be pollinated by birds. In the open canopy of the gymnosperm forests most pollination could happen by wind. Descendants of the pines or the firs like the monkey-puzzle, the cryptomeria, or the giant sequoias saturated the air with pollen, and no other vector was needed. As the angiosperms diversified and forests thickened and developed 50 million years ago this was no longer possible, and a new strategy for pollen dispersal was needed. The solution was the insects, but also the birds, the 11,000 species of animal that by now were navigating through the trees to find their food.

Birds, like dinosaurs, have an eye structure that can instantly spot the difference between red and green. Dedicated scents in flowers are generally developed to attract insects, whereas other pollinators like birds tend to smell the nectar itself, which is why red tubular flowers are unlikely to smell. For a fruit to stand out against the green it became red, and to promote seed dispersal at the right time sugars were wrapped in carbohydrate coverings, which would turn bitter to sweet as the seed cases hardened. And just as with the cycads, seeds developed a range of poisons as well as a hard casing to ensure they stayed intact in the bird's gizzard. Most of the time these poisons, like those employed by the cycads, are glucosides, simple, lethal molecules bound to glucose or another sugar. It's a genius system: plants produce sugar, so are unlikely to break these molecules down, as they tend to be building chains of molecules up, but an animal's digestive system will snip the bond immobilizing the poison and thereby destroy itself. In apricot, apple, and plum the glycoside is amygdalin, and when its sugar is snipped off the tiny amount of cyanide is released. Cyanide is lethal to humans but an effective poison across the animal kingdom, a carbon and a nitrogen that stop energy flow at the most fundamental level. In the horse chestnut esculin is the poison in question; snip off the glucose attached to it and you get coumarin, the smell of tonka bean and vanilla and an appetite suppressant, designed to stop the grazers biting. The cycasin of the cycad is one of the most lethal of all, releasing formaldehyde to pickle its victims and diazomethane to knock them flat.

The velvet glove and the iron fist—poisonous seed and soft, delicious fruit—have endured as methods of harnessing birds to move seeds all over the world, from the equatorial Amazon to the temperate rainforests of Turkey.

From Çamlıhemşin near the Georgian border I walked up the mountain toward Pokut Yaylası, the traditional Hemşin grazing area at the head of the valley. The valley sides are so steep here that zip-wires are used to send food up the mountains and across the valleys. The first snows had come early, and cloud was slowly drifting up the valley. A damp pearling became a light drizzle and finally a long, fresh rain in whirling mist.

The deep valleys of the Pontic Alps encourage trees to grow that have huge leaves and drop tips, adapted to lots of rain and very little wind. They include the cherry laurel and the Caucasian holly, the Caucasian buckthorn (*Rhamnus imeretina*), which can have leaves 10 inches long, and the Pontic oak, *Quercus pontica*, which has huge oval leaves. As I went up the hill I was struck by the variety and vibrancy of the berries. The holly and the viburnum had berries of similar intense red, the leaves were totally gone from the rowans and only the berries remained, hanging in clusters from the trees. Little birds were everywhere: coal tits and thrushes, sparrows and warblers. Due to the lack of wind, many of the trees, particularly those that exist in the understory, have evolved red berries that will be spread by birds. Bears, ambling with their airy fluidity through the forest, also enjoy the nuts and berries once they have fallen to the ground, but they are less common. The real target of all these berries—the holly, the rowan, the *Viburnum*, and the *Vaccinium*—are birds.

"What wondrous life is this I lead / Ripe apples drop about my head . . . The nectarine and curious peach / Into my hands themselves do reach!" Andrew Marvell's poem is gently whimsical, but it is also literally true. Ripening time is nothing more than the tree equivalent of calling a taxi. Apples, for example, will drop when ethylene gas stimulates the production of cellulase and polygalacturonase, both enzymes that cut the glues that hold the cells of the stalk together and neatly sever it, depositing the ripe fruits on the ground for their distributors. This is another way trees manipulate animals, to the point of sometimes literally controlling their life cycle. Oddly enough, the dodo and other flightless birds are superb examples of how trees have done this.

The name *dodo* comes from *"dodaerse"*—in Dutch, literally "fat-arse"—a name given to it because when these huge birds were first seen by Dutch sailors in the seventeenth century they waddled slowly around with their bottoms in the air.[5] It was June, so the dodos were in their furry summer plumage, with a curl of feathers above their protuberant bottoms. Had it been January, then the sailors who landed on Mauritius would have seen a different bird entirely: slim, agile, and so lean that sailors who ate it had to boil it for hours and even then complained that the tough, leathery flesh was nauseating. This fat-thin cycle was shaped by the trees of Mauritius and the timings at which it was advantageous for them to drop their fruits. It may be that the closed ecosystem of the island even allowed for a measure of cooperative coevolution between the trees, which kept the dodo flightless. The dodo was forced to fast or eat insects and small plants through the rainy season from December to March while the trees were flowering. In April the palms and pandanus trees started to drop their fruits by the shore and the dodos would gorge on these until October, when they would retire to the forest and eat fruits dropped from the forest canopy. These were scarcer and came to an end in December, when the fasting season would begin again. This fast period ensured, counterintuitively, that the dodo remained large and flightless. Only a large bird could survive without food, and because the dodo was flightless it had no way of eating the fruits before they were ripe. It was important for the trees to choose a large bird, because of the size of the seeds that could safely traverse its gut.

Until humans came along the dodo didn't need to fly. The trees dropped their highly calorific fruit on the ground for it at the perfect time, and it had no natural predators. With humans and their vile gluttony came rats and other predators, and the dodo quickly became extinct. So too did many of the trees of Mauritius, and while epic deforestation—as much as 95 percent of the island—is explanation enough, some trees were thought to have become very rare or even extinct because their distribution technique, the dodo, no longer existed.* One was the tambalacoque (*Sideroxylon grandiflorum*), a tall sapote tree that in 1973 was thought to survive in a population of just thirteen trees. Its fruits, a little like peaches, have big, tough seeds that need to be scraped before they can germinate, and the stone-filled gizzard of a dodo, the theory went, provided the necessary shearing tool. Attempts to prove this by force-feeding tambalacoque fruit to turkeys were peculiarly unenlightening, however, as the turkeys tended to spit the fruit back out.

The theory became one of the great scientific spats of the 1970s and went on to join a debate that enlivened the scientific world throughout the 1980s: to what extent were many of the fruits eaten by humans actually designed for consumption by extinct megafauna?[6,7] The golden age of the megafauna came in the early to mid Pleistocene; since then they have been in decline, perhaps as a result of glaciation. As with the dodo but 15,000 years earlier, the spread of humans around the globe was the final coup de grâce for most of them; as K-strategists—living long and investing much into their offspring—they could not replace those who were killed.

The biggest mammal to survive the extinction event 66 million years ago weighed less than a pound, but less than 100,000 years down the line their descendants weighed 13 pounds, and 200,000 years after the asteroid they weighed 44 pounds, far heavier than any mammal that had existed before the asteroid. About 15 million years ago the true megafauna appeared, reaching their apogee around 2.5 million years ago. There's little doubt that trees shaped the megafauna, feeding them fats and proteins and creating the perfect evolutionary partners for spreading big seeds. Such seeds are essential for establishing new trees under an already existing forest canopy, as the seed must have plenty of opportunity to reach the light, and so concocting guava fruit full of protein or avocados full of fat was a way of both attracting the megafauna and making sure they stayed large—large enough to carry an avocado pit across the Amazon Basin undigested.

Careful to attract megafauna and nothing else, avocados don't

---

* Another of the trees that is desperately endangered on Mauritius is *Dombeya mauritiana*, a very beautiful type of large mallow tree.

change color as they ripen, nor will they ripen while they are still attached to the tree. Their rich flesh, full of incredibly dense nutrients, was designed to attract gomphotheres, the now-extinct giant mammal that roamed across South America until about 20,000 years ago. The pawpaw and the persimmon, the jackfruit and the jumari fruit, the mangosteen and the ice-cream bean all appear to have evolved to drop at the feet of the megafauna and be gorged on before they sloped off to another tree many miles away. Even giant sloths appear to have done their bit, ensuring that the Joshua tree (*Yucca brevifolia*) was able to hover on the edge of a desert but be widely dispersed enough to survive climactic change.★

As a collective, trees used variable fruiting times to keep megafauna on the move. In the same way that bird- or bat-pollinated flowers spread their flowering times out, so the fruiting of trees dependent on megafauna could ensure movement around the forest. Attempts to map these routes have been made with variable success, showing the way an animal's distribution allowed for the genetic spread and distribution of a certain tree. Flipping the axiom would show that the trees that we think of as large and stationary can be agile and fast, spreading across land before retreating back again in response to climate and landscape and enlisting and shaping different species of animal in order to do so.

Examples of trees using animals to move can be seen across the animal kingdom, but there's often some ambiguity about who is shap-

★ Preserved sloth dung has been found containing Joshua tree leaves and seeds, confirming that they fed on the trees. During the Pleistocene epoch that lasted from about 2,580,000 to 11,700 years ago, *Nothrotheriops shastensis*, or the Shasta ground sloth, roamed North America in what is now central Mexico to the southern United States. *Nothrotheriops shastensis* was one of the smallest species of ground sloth at 9 feet from snout to tail and weighing nearly 400 pounds! The animal had large hind legs and a muscular tail that would be used to support the animal when it wanted to stand on them. https://www.lancastermoah.org/single-post/pollinators -of-the-joshua-tree-past-and-present

ing whom. Darwin's favorite example of a tightly evolved relationship between a tree and an animal was the pollination system of the Joshua tree, which in a letter to Joseph Hooker, then director of Kew Gardens, in 1874 he called, "the most remarkable fertilisation system ever described."[8] It relies on the tree exuding a scent (the chemical nerolidol) that reminds the small brain of a highly specialized moth to lay its eggs, and as such it's an example of extraordinarily tight coevolution, as well as an extraordinary perversion of the usual herbivore-tree interaction.

The Joshua tree is a desert plant with fearsomely spiky leaves and waxy green flowers that face the sky. The flowers are pungent, and the chemicals they pump out are particularly attractive to both males and females of a small brown moth species, the yucca moth, which fly to an anther, gather the sticky pollen with bizarre mouth-tentacles, and place it on the stigma of flowers from other trees. The moths then sit on the stigma and use an ovipositor the same length as the style (the stalk that links stigma and ovary) to lay their eggs just next to the seeds. Fertilized seeds and eggs codevelop in what is essentially the same womb until the caterpillars hatch, whereupon they eat a few, but not all, of the seeds—their only food source—and hurry off to pupate somewhere nearby. The Joshua tree does not provide any nectar or other reward, but genetics reveal that this surrogacy partnership has been going on for millions of years.

Why would the flashy, iconic Joshua tree bind itself to the fate of one fragile moth? The advantage for the tree is absolute agency over what its pollinators will do. It has no need to produce nectar, which might attract animals that could harm the tree, and the energy it offers the moth in terms of fats in the seeds is generous but delayed, like a consultant's salary. The tree is the primary producer of energy, and in that sense holds all the cards, but even more, the absolute specificity of the female's mouthpiece is a guarantee that the moth relies on the tree's survival. Moreover, the tree will abort flowers if too many of the seeds have been eaten, killing all the seeds/larvae inside and ensuring that the larvae will never start to feast too soon.*

---

*Major extinction events and rapid climate change tend to lead to powerful coevolution, both to ensure longevity and to ensure proper gene mixing in small

Last year I went to the Mojave Desert in California, where the Joshua tree grows, to investigate a twist in the tale of what remains one of the first and strongest examples of evolution by natural selection. Darwin knew about the Joshua tree through the work of his correspondent Charles Valentine Riley, the state entomologist of Missouri, who was able to envisage how plant-animal coevolutions might work better than anyone else in the scientific establishment. That said, Riley is better known as the savior of the Old World wines and was given the Légion d'honneur for what was essentially his services to coevolution: understanding that American (*Vitis*) vines, having coevolved with the *Phylloxera* aphids that were annihilating the vineyards of Burgundy and Bordeaux, would be resistant to them. By enabling vineyards to replace their ravaged rootstocks he saved the French wine industry from almost certain annihilation.

Riley applied the same insights to the Joshua tree. When he got to work in 1872, the tree was already iconic to American settlers. Its spiky, awkward silhouette in the desert was like a thin old man trudging westward, and so Mormon pioneers named them after the Old Testament patriarch who led the tribes of Israel into Canaan. Riley was asked by someone who had seen small brown moths in the large flower spikes of the *Yucca* to investigate, and so he did. For the next ten years he fanatically observed the life cycle of the yucca moth (*Tegeticula yuccasella*) and the bogus yucca moth (*Prodoxus decipiens*), and accurately documented their life cycle. He used his observations to contribute to debates on mimicry and animal pollination and sent the moths in their cocoons to other scientists. Darwin promptly sent his on to Hooker to look after at Kew Gardens, a transfer of needy scientific specimens that continues to this day. Amazingly, at least to us, some of Riley's observations were refuted by people who believed that active pollination "belongs in the land of fables." It was not so many years since Christian Konrad Sprengel, the first person in Western science to postulate that flowers were designed for anything

---

populations. Cycads, for example, tend to use specific beetle species as pollinators to ensure that one animal has to go from tree to tree, as their populations have become increasingly isolated.

other than to be beautiful, had put forward his ideas that insects were "living brushes," and of course still before Darwin's idea of evolution by natural selection had become widely accepted.

I drove out from Los Angeles at the end of October, when the nights were cold and the last seed pods were just falling off the trees. Since 1872, scientists have discovered that the Joshua tree is not one but two subspecies, *Yucca brevifolia brevifolia* and *Yucca brevifolia jaegeriana*,[9] each exclusively pollinated by separate sister species of yucca moths with obligate seed-feeding larvae. Controversy now swirls around whether it is the moths themselves that have driven these two subspecies apart, or whether their climate niches forced the moths to coadapt so they became sisters rather than one species.[10] I wanted to see what this important example of coevolution between tree and pollinator actually looked like, so I drove along the edge of their range, through the Joshua Tree National Park and then east along the edge of the Mojave Desert, until my mind was full of angular trees, rank after rank, dancing like scarecrows across the desert.

The difference between the subspecies was surprisingly easy to see. In Joshua Tree National Park the trees were tall and proud and gaunt, with long sweeping spikes of leaves and few branches, dry seed-pods, and occasionally a dead branch languishing on the ground. As I left the park and drove east, I passed rank on rank of Joshua trees, then occasional patches, until as I continued east from Barstow, rejoicing in the romance of the straight California roads and the road signs to Las Vegas, I started to see a new form. At first it was just one or two; past the hybrids of the Tickaboo Valley and there was a profusion of squatter, more industrious, and prosperous-looking trees, with many more branches and nodding pom-poms at the ends of short leaves. Whereas *Yucca brevifolia brevifolia* branches only after it has flowered, *Yucca brevifolia jaegeriana* branches quickly and often, hunkering down to the ground. It's one of the peculiarities of Joshua trees that the ends of their branches have to be nipped by frost before they will flower.

I never made it to Vegas, but the Joshua trees do. In fact, remote sensing of the range using LiDAR has shown that Las Vegas is completely surrounded by *Yucca brevifolia jaegeriana*. Indeed, it can survive in areas that are even drier and hotter than its variant. Analysis of the moths suggests the species divided 10.7 million years ago, but both the sisters and the two Joshua tree subspecies they rely on remain extraordinarily similar. Both types of Joshua tree have flowers that open to the south and pick up the maximum amount of solar energy, for example—a way of providing a thermal reward to a moth pollinator that only emerges after the sun has set and of warming up and sending out the scent to attract their pollinator like a solar-powered scented candle.

The scent is variously described as blue cheese, musky, or slightly rotten lily. For me its cloying fragrance is closest to the smell of a Swiss restaurant, a Gruyère fondue. In 2016 a beautiful experiment in the Tikaboo Valley defined the chemicals that produced this all-important profile of scents. The researchers put oven bags over the flowers of each species and the hybrids they found growing near them and captured the scent the flowers were giving off. In a mixture of twenty different chemicals there was one clear and unusual base note: the eight-carbon alcohol 1-Octen-3-ol. The head notes were unique: alcohol and lactone derivatives of (E)-4, 8-dimethyl-1,3,7-nonatriene (DMNT), including nerolidol, which smells like a breath of fresh bark. These chemicals are of interest because while the first is known as mushroom alcohol and is found in human sweat and breath, where it acts as an attraction mechanism for hungry mosquitoes, the second is normally released by damaged plants. While no explanation is needed for Joshua tree flowers smelling like human BO (a huge number of biochemical pathways are shared between all branches of the tree of life), it's likely that the signals for damaged cells play into a pathway that causes the moth to recollect how to use its ovipositor. What probably drove this speciation is the length of the ovipositor: wounding of ovules during oviposition prompts flowers to be aborted, and so the ovipositor length must exactly match the distance into the ovary.[11]

What this means is not only that the Joshua tree is prompting the moth to lay its eggs in the moment by the simple use of a scent, but also that over millions of years it has literally shaped the ovipositor of two moths to the millimeter. It's both extraordinary and intriguingly pointless—why would a tree evolve to find itself in this position? The answer is probably that it's extremely efficient, energetically very cheap, and extremely predictable and effective. The moth is the IKEA of pollinators, not traveling very far, but far enough for the clustered populations of the two trees, and providing a reliable if unspectacular service time after time.

Under a gloriously blue afternoon sky in the Joshua Tree National Park I followed a trail to some petroglyphs, 2,000 years old, carved by American Indigenous people as they traveled through. The twirls and angular patterns were beautiful, absorbing, and strangely relatable, although I could only begin to guess at what they meant. I longed for someone to ask, but the last two thousand years have wiped out all the people with an inkling. With the cold, dry air nibbling at my lips I climbed the rock and sat down to look across the landscape, with the lengthening shadows of the Joshua trees marching across it. Shaggy and angular, they looked a little like giant sloths themselves.

Two thousand years back in time felt not so long. Twenty thousand years ago mammoths, camels, and saber-toothed cats had wandered over that landscape when it was wetter and greener. Perhaps 12,000 years ago a giant sloth might have lumbered gently up to one of the trees and, pressing its tail firmly into the ground and sitting up on its hind legs, munched hungrily at the fruits on the ends of the branches. It may even have sniffed hungrily at the flowers' mushroom-alcohol scent. Suddenly I thought back to the moths that lived in the hair of the three-toed sloth, and its slow descent down the tree: Was the giant sloth in its turn covered with Joshua tree moths? While the giant sloth roamed mile after mile over these rolling hills with the fruits of Joshua trees gently digesting in its gut, skirting round the granite extrusions, stopping to drink at the occasional pool, did the moth cling onto its hair, guided there by the volatile scent of body odor, the mushroom alcohol 1-Octen-3-ol? Did defecation of the seeds then prompt the moth to drop off the giant sloth and burrow in, laying its eggs next to the seeds? And by the time it had completed the larval stage of its life cycle, was the Joshua tree growing up, capable of being pollinated so that the cycle could begin again?

As the sun set and stars started to shine with desert intensity, my imaginings became more excited . . . When the giant sloth stopped visiting the tree and eventually became extinct, did the stranded moths fly instead to the half-remembered smell of a wounded leaf and a hint of mushroom alcohol and lay their eggs in the tree rather than the sloth's dung? And could the tree have resurrected some sort of crypto-gene that had made mushroom alcohol in a shared ancestor billions of years ago to manipulate the moth? Or had the whirring biochemical pathways of the tree experimented with chemical after chemical, bouquet after bouquet, until one had done the trick and attracted the moth every time? As with the meaning of the petroglyphs, the person who would be able to tell me what was going on was probably not walking the earth, but, like Foucault after his weekend in California, I felt like crying and declaring that I knew Truth.

It was pure supposition, but no stranger than many of the routes by which trees have developed to shape animals. A unique chemical pollination signal is essential for many trees. The dry, dusky smell of fig leaves is there to enhance the smell of the fig flower, so that isolated trees in rocky landscapes can attract wasps from far away.[12] The inside of a fig is full of unisex male flowers, and specific female wasps are attracted to each species of fig, pollinating the flowers and laying some of their eggs inside. The larvae feed on the developing ovules and emerge into the central cavity of the fig when the fig's male flowers are mature. The wasps then load themselves with pollen and buzz off to search for a new receptive fig of the same species.

In all of this, the scent the tree is producing is complex. Not only does each species have a slightly different bouquet, but this bouquet changes over time, so that male and female figs smell the same whenever it is essential that their wasp pollinators cannot distinguish between them. After pollination the fig's floral scent changes to keep wasps away from already pollinated figs.[13] Even though Mediterranean figs were domesticated around 5500 BC, if not before, they maintain this floral signal, and farmers have known the outline of the pollination strategy for much of that time.* Trees will also "spike"

---

* Scent is an extraordinarily important conduit of information between trees and animals, and one that scientists think evolved very early on, and something that we have the remnants of ourselves. Producing a scent is energetically costly, and also risky as it can (and does) often attract herbivores. Sometimes a pheromone is the

their nectar with mind-altering substances to attract insects. Bees are an example of animals that trap-line forage: they have a 2D mind map with all the sources of nectar and pollen marked on it that they will follow time and time again. If trees put a psychoactive substance into their nectar they become more strongly marked on that 2D map, ensuring that the bees return. Sometimes this can backfire: dead bees are often found under lime trees (*Tilia tomentosa*), for example. This is because limes can put a caffeine-like substance into their nectar that stimulates bees to the extent that they forget to look for other sources of sugar and persist in returning time and again, desperate to get their hit even when the flowers have no nectar left to give.

Pollination and seed dispersal are two major reasons why trees shape animals, but the most universal is herbivory. At the beginning of this chapter we met the three-toed sloth, a relentless and deliberate herbivore, eating the leaves of cecropia, one of the most widespread and successful trees in the tropical rainforests of South America. Aphids and most other herbivores avoid cecropia, because the tree is protected by *Azteca* ants. Not only does the tree provide the ants with carbohydrates, in the form of glycogen-containing structures at the bottom of its leaves and particularly the young ones, but it also protects them with a nest consisting of multiple chambers within the stems. And should any vines attempt to climb the tree, posing the risk of pulling it down altogether, the ants will cut through them.

This sort of relationship is called defense mutualism: in return for the protection of the tree, the ant will defend it to the death. Many tropical trees enter into such relationships with ants, sometimes extending three ways and sometimes weaponized against other plants. One of the most striking examples in the rainforest is the cockspur tree (*Vachellia cookii*), which has massive red thorns that protect the ants inside them. In return the ants attack predators that touch the

---

scent used, and pseudocopulation is shown more and more often to play a role in pollination. https://www.nature.com/articles/s41598-023-32450-6

tree. Another acacia (*Vachellia gentlei*), the red cockspur, houses ants that release pheromones to warn the herbivores before rushing out and attacking them. The tree feeds them on Beltian bodies, protein nodules that grow on the tips of the leaves, and houses them in thorns. A tree, *Zeyheria montana*, from Minas Gerais (a much drier part of Brazil) recruits two insect partners, as well as being pollinated by hummingbirds. It attracts patrolling ants, like the aggressive *Ectatomma tuberculatum*, using extra-floral nectaries, pores that release small amounts of sugar, and these in turn defend the leaves against herbivores.[14] In addition, the reproductive tissues of this shrub host the treehopper *Guayaquila xiphias*, which provides honeydew to *E. tuberculatum* in exchange for protection. This trophobiotic relationship persists throughout the plant's reproductive period and seems to be effective not only as a defensive strategy for flower buds and flowers but also for the fruit, which, although dry and dispersed by wind, is attacked by weevils because it has absorbed a significant amount of water during its development.

Finally, the tree *Duroia hirsuta* uses the help of the lemon ant to construct what are called devil's gardens, in which only a few species can survive. *Duroia hirsuta* is allelopathic, pumping out the strong plant growth inhibitors plumericin and duroin, which are both similar to iridomyrmecin, a defensive chemical classified as an iridoid isolated from ants of the genus *Iridomyrmex*. In wasps such as *Leptopilina* it has also evolved into a sex pheromone, with host species using the smell of iridomyrmecin to detect the presence of the parasitoid wasps. Among the variety of plants iridomyrmecin is found in is *Actinidia polygama*, whose slightly swollen stem is hollow, the perfect place for a queen to install herself. Once the ant colony is established it destroys the other trees around the host, cutting them and spraying them with formic acid. The tree grows high and spreads, and new ant colonies will eventually colonize its offspring.

The species we see making up the earth's flora and fauna are tiny, lucky fragments of what might have been. Today, we have the hornbeam, *Carpinus betulus*, but we only know about its close relative *Carpinus perryae* from a single slab of rock found in Washington State. We have the Scots pine, *Pinus sylvestris*, but all we have of *Pinus yorkshirensis* are a few microscope slides and a single image, because the only fossil specimen, which was dug out of the Speeton clay deposits in Yorkshire, was destroyed in the process of establishing its age: 130 million years. And we have *Ginkgo biloba*, conserved by Chinese monks when all its relatives, including the beautiful *Ginkgo gardneri*, petrified in nineteenth-century Gothic in the Natural History Museum in London, died out.

Outside my house in London three beautiful young *Ginkgo biloba* specimens had been planted in the pavement by the Westminster tree officer, and every year I watched as they sprouted little axe-shaped leaves of bright parrot-green all the way round its twigs, which then broadened out into great fan-clad arms, and then in the autumn the leaves turned canary yellow before dropping a sheet of gold onto the street. They were wonderfully strange, lacking so much of the directionality we assume (extrapolating from the evidence of oaks and beeches and chestnuts) all trees will automatically adopt. Fossil evidence shows us that ginkgo is one of the survivors, a tree that since the early Jurassic period has changed very little for 200 million years.[15] In the wild, ginkgo appears to have been outcompeted by the ability of angiosperms to grow back after serious disturbance. Notwithstanding this, it is chemically very tough, able to resist a broad spectrum of pathogens and herbivores. And almost miraculously, its response to a caterpillar munching through its leaf is nearly identical to the response of an angiosperm, which might have evolved 150 million years after it. Ginkgo expends a lot of its energy on preventative compounds: strange, gluey ring chemicals that make sure many herbivores never even start to nibble on the leaf. If they do bite more than once, the protein floodgates open. Calcium floods through the leaf, changing the membrane potential—the charges across membranes—to a far greater extent than simple mechanical damage. A zinc-finger protein—a loop of protein around a central nail of zinc—encourages upregulation of certain genes and more proteins, producing poisonous chemicals. Sesquiterpenes, terpenoids, shikimic intermediates are all churned out. Some of the chemicals drift away on the wind, invoking defense responses in the leaves they diffuse into.

"It doesn't matter whether you compare kangaroos, bacteria, humans, or salamanders," said Lynn Margulis, "we all have incredible chemical similarities." To this we might add that we have incredible similarities with trees.[16]

In short, although there are many tiny, shuddering variations in the ways trees respond to being bitten, the broad outline has been going strong for 200 million years, and it includes emitting volatile organic compounds. This is hugely significant, because it means there's a basis for tree-animal interactions that is almost impossibly old, older than the hills, older than the continents, as old as dinosaurs, and trees communicating damage to other trees (or perhaps a handy predator) is part of that ancient interaction.

So why and how have trees shaped animals, when it could have been the other way round? The answer can be seen in a collection of fossils in Germany, the famous Messel pit. Forty-seven million years ago it was a lake surrounded by subtropical rainforest, the new multilayered broadleaved forest that had taken over after the Chicxulub asteroid had annihilated the dinosaurs and dislodged the conifers. Global warming over 10 million years had allowed vegetation to boom, and the overall effect would have been not unlike what happens on the shores of the Amazon. When carbon dioxide seeped over the lake and into the surrounding forest around it, it killed a selection of animals that would not look completely out of place in the Brazilian rainforest today.

Their perfect preservation shows that animals—particularly mammals—evolved in response to the rise of angiosperms and the encroaching jungle in a remarkably short slice of evolutionary time. Among the fossils is a bat, which shows evidence of flight and flaps of skin—large ears—that show it has quickly evolved to echolocate its way through dense forest. Another rapidly evolving mammal has developed an opposable thumb, one of the first examples of an animal that had evolved to climb trees. Assessing the two speeds of evolutionary change against each other shows that the primary producers—the trees in this situation—were driving the changes.

The trees were essentially in a lower gear, driving the change forward, while animals in a higher gear responded to it physically with remarkable (and remarkably quick) plasticity.

But although animals physically accommodate themselves to trees, trees are devious in the substances they produce. Many of these substances mimic a hormone, block an element of the central nervous system, or, like nutmeg, suffocate the animal to death at a cellular level. Few are targeting humans or even primates specifically: they are blanket-poisoning herbivores. Targeting, however, is exactly what *Theobroma cacao*, the cocoa tree, did, using theobromine, or food of the gods, the active ingredient and distinctive taste in chocolate. The tree was isolated in a kidney-shaped niche around Ecuador during the last Ice Age about 21,000 years ago,[17] but after the world started to heat up the tree spread out of its refuge astonishingly quickly, becoming a crop plant across South America and a form of currency for the Mayan civilization. This happened because theobromine, like cocaine, caffeine, and nicotine, is a ring alkaloid, a ring of carbon and nitrogen that fits neatly into adenosine nerve receptors. It has a mild but pleasurable effect on the nervous system that makes it irresistible to primates. They get, to a greater or lesser extent, a psychoactive high, which allows them to swing through the canopy for miles.

This is essential for a tree that can only grow in specific niches that are scattered about the forest and likely to be a long distance from one another. Ruthlessly, the trees select for larger monkeys able to travel further distances by using the toxicity of theobromine: just as chocolate can kill dogs, so too large a dose would knock off a small monkey. To ensure that bigger animals can eat the seeds, the trees are monoecious, producing separate male and female flowers on the same tree, and the female flowers grow directly on the trunk and larger branches, rather than on the ends of delicate new shoots, by a process called cauliflory. The flowers are pollinated by midges, and large, heavy pods can then develop without breaking branches on the parts of the tree that can support the weight of larger monkeys.[18] These large monkeys can eat the pods and turn the very high fat content of cocoa butter into the physical energy required to travel long distances, and the theobromine hit into mental energy and spread the seeds.

The use of theobromine as a drug to encourage monkeys to travel harder and faster seems a world away from horse chestnuts fluttering their white flowers and using ultraviolet patterns to attract bees to their pollen. In both cases, however, the tree is using its extraordinary powers of chemical synthesis to manipulate the animals it will use to travel. It's like a car factory producing machines of various designs that enable people to travel distances infinitely further than they could alone, but it's a design process that started out of shared basics of life and biochemistry. Chemically, physically, and existentially, animals have been shaped by trees, and (with the possible exception of the sloth) no animal has been more shaped by trees than humans ourselves.

CHAPTER 8

Trees shaping people

How trees shaped humans physically, mentally, and politically, through interactions with earth, wind, and fire

Trees are an invitation to think about time and to travel in it the way they do, by standing still and reaching out and down.

Rebecca Solnit

Once upon a time there was a girl who lived in a tree. She had deep-set brown eyes and brown hair. She ate fruit—orange mangosteen and black juniper berries—crunched on nuts, sucked on sweet grasses, chewed juicy leaves from the ground, and dug up tubers and roots, knowing which ones were good and which ones were hard or poisonous. Sometimes she followed the trails that criss-crossed among the trees and through the grass, but much of her time she spent clambering through the broad crowns of the trees that grew across the savanna, reaching up for branches and feeling the texture of the bark against her hands, balancing against the trunks and springing along boughs, bracing herself against another branch above her head.[1] At night she tucked herself into the fork of several branches and curled up to sleep, watching stars like diamonds and the black of the branches against the sky.

One day, high up in the canopy, she slipped. She fell more than 40 feet, grabbing at leaves as she fell, but nothing broke her fall and she hit the ground at over 34 mph, feet first, then fell forward onto her outstretched arms. The impact was too much, and her legs and pelvis shattered, as well as her arms and ribs. There was no medical help, and she died from her injuries in a couple of hours. Then, 3.2 million years later in what is now Ethiopia, her fossilized bones were taken into an archaeological camp where they were playing a seven-year-old Beatles song on repeat, "Lucy in the Sky with Diamonds."

By 2016 Lucy One, as she was now called,⋆ was an icon and made a posthumous tour of American museums from her base in the museum in Addis Ababa. As she passed the University of Texas, John Kappelman and his coworkers used an MRI machine to scan her bones and reconstructed the breaks in them with the help of a trauma pathologist.[2] They found that Lucy had complex compression fractures in her larger bones, and the smaller ones had broken in a "greenstick" manner, a half-break, half-split often recorded in pathologists' reports on children who have fallen out of trees. They put together a picture of how she might have fallen, and thus a famous fossil came alive. Some scientists have dismissed this as a just-so story,[3] pointing to the fact that a fossil 3.2 million years old inevitably has breaks in nearly all its bones—it has been buried, traveled far under the compression of rock, and then suffered all the wear and tear of reemerging. So we can't be certain about her death. But we can be pretty certain that she lived her life in trees.[4]

Lucy was famous because she was the first hominin who was known to have walked upright and straight-legged like a modern human.[5] Her species, called *Australopithecus afarensis* after the Afar area of Ethiopia where she was found, had adapted to walk upright for a good distance. Her kneecaps sat as they would on us, her pelvis was a similar shape to that of a modern woman, and she had the lumbar curve at the bottom of her spine that gives our backs their improbable and beautiful shape. Generally, *Australopithecus* had exchanged the clinging, wrapping feet of tree climbers for high, stiff arches that help with walking, although Lucy herself had asymptomatic flat feet.[6] However, she also had hefty shoulders and strong arms and, presumably because of the risk of predators such as saber-toothed lions, she was almost certainly still sleeping in trees and using them to move around when, at some point before her eighteenth birthday, she slipped.

⋆ Or in Amharic, "*Dink'inesh*," which literally translates as "You are marvelous."

Human history, the human brain, the human hand, and the human leg, all start in the forest canopy. Our ability to bounce along branches standing upright, our desire to build nests and smell cedarwood— all the adaptations of primates—are adaptations to trees. A tree is a knotty place to live: three dimensions of complex branching, a trunk that bifurcates into successively more perilous and delicate boughs, branches, and twigs before producing leaves. Binocular vision, with both our eyes pointing forward, allows us to judge distances and see and map a way through the canopy. Standing upright you can reach a branch high above your head and grip it with strong and flex-ible hands that have the perfect adaptation to smooth bark. Fingers backed by hard nails have fat and fluid-filled pads surrounded by a stiffer lining, allowing them to deform like a slightly deflated tire to the shape of the branch and maximize surface area and resultant fric-tion. Fingerprints, which we share via convergent evolution with the otherwise very distantly related koalas, channel away a film of water that might cover a branch and interlock with rougher bark. They also allow for extreme sensitivity by magnifying strains and allow us to avoid blisters. Claws were no longer necessary and evolved away.

All these changes were in place in tiny primates (rather like bush babies) by 50 million years ago, but these primates weighed less than a pound, and their brains were minuscule with, we can safely assume, a correspondingly diminutive neocortex, the gray matter that deals with higher-level thinking. In bush babies today the neocortex oc-cupies 47 percent of the brain, considerably less than the 76 percent it occupies in a chimpanzee and 80 percent in humans, and by looking across the range of primates alive today we can see that body size, neocortical size, and intelligence are all linked.

Trees and tree-dwelling shaped the physical evolution of the early humans. Just like other animals that made the switch from eating leaves or insects to eating fruit, the primates were deeply affected by their different diets. Some leaf-eating monkeys still exist, notably the proboscis monkeys of Borneo, known for their long pink noses and their stomachs, distended from eating quantities of leaves and slowly digesting them. Leaves and shoots are easy to find, but full of cellulose and lignin, as well as tannins and other tree-made poisons that make them hard to digest, so a leaf-eating primate must eat large amounts of young tree-growth and then hold it in its stomach to digest and detox it. The result is that proboscis monkeys migrate around gently in small groups, often through mangrove swamps, and have slow metabolisms and a peaceable nature and are not very intelligent for their size. They are an exception among primates, most of which have few leaves in their diet, and are particularly threatened by deforestation.

The insect-eating primates, by contrast, are the small lorises and bush babies. Catching insects is a tricky, energy-expending business, and so the species have remained small and reactive, unable to grow much bigger on an insectivorous diet. Their brain size remains small. The way around the energy expenditure problem is to hunt down and eat fruit. Fruit is full of fats and oils and sugar, a perfect package of energy, and it doesn't run away, so primates can grow large very easily. The problem is that any primate evolving in this direction has to remember where each fruit is and what sort of sequence it operates in. Even fruits of the same species may become ripe at different times. Fruit-eating primates, in other words, find themselves in thrall to the tree, forced to be there when the tree produces the fruit, not at any other time. The result is that fruit-eaters develop intelligence and memory in the form of spatial and temporal awareness. They're able not just to find the tree but remember it. Spider monkeys, for example, have brains on average 25 percent bigger than leaf-eaters and can remember large numbers of fruiting trees, the times that they fruit, and the most efficient route between them. As a result, they're superb seed spreaders.

Our closest relatives, the orangutans, bonobos, gorillas, and chimpanzees, have brains that are huge for their size.* In Borneo, the Dayak people have a legend that orangutans could talk if they wanted to, but prefer to remain silent so they're not forced to work. Research into their brains shows that they haven't actually developed for speech—the need for a large brain seems to have come instead

---

* And in the case of orangutans, at least, it certainly isn't all being spent on societal manipulation.

from the need to have a sense of bodily self-awareness. This allows orangutans to move slowly through the canopy with their weight spread out and to adjust their movements according to their level of peril—walking on all fours above or below the branch when a branch is thick and large or spread-eagling across smaller branches to spread their weight out as much as possible. Their huge brains allow them to minutely judge the flexibility of trees, bending smaller ones back and forward to trebuchet themselves across from tree to tree.

Susannah Thorpe, the researcher who discovered this, also realized that the tendency of orangutans to stand upright on flexible branches and clutch other branches above their heads meant they used the springiness of wood to propel them along, rather like athletes with prosthetic running blades or children bouncing on trampolines. This springiness and flexibility in branches is the result of the long tracheid cells and the xylem and phloem tubes that run up and down trunk and branches longitudinally; cells with crystalline microfibrils of cellulose coiling around them, embedded in a matrix of hemicellulose and stabilized by lignin, an organization that enables boughs to bend and not break. Whereas the first trees to develop in the Devonian period, 380 million years ago, snapped as soon as they were hit by water or a landslide, modern trees tend to bend and spring back.

A wonderful example of this is the European beech, which in the Carpathian Mountains can be flattened by snow until it lies horizontally. In early spring a few years ago I walked up the Creasta Cocosului, the mountain that looks like a cock's comb. The snow was still deep on the top of the mountain, melting in patches, with delicate purple crocuses coming up through it and some dried and golden grass. On the slopes next to the summit large fir trees were growing, but below them were great sheets of young beech trees that had been pushed down over and over again by the snow, creating a strange, brown, woven surface. I walked down the mountain on one of these mats of young trunks and, apart from the occasional moment when I slipped painfully through the branches, it was an exhilarating experience. Bounding down the mountain on springs 5 feet above the ground felt superhuman and surprisingly relaxing. For a hungry orangutan rushing to catch the seasonal fruit from a tree 5 miles away wood's pliancy makes a natural springboard. The sprung wooden planks of a ballroom floor are designed for the same effect: powered by gravitational energy stored in the elastic grain of wood you can dance all night.

During these long rambles through the treetops orangutans developed a "feel" for wood, just as humans do. The stiffness and flexibility of wood vary enormously from tree species to tree species, but it's more than that. Spend any time in trees and you can feel their different chemistry, slipperiness, wood density, the angle of branch to trunk and branch to ground, the texture of the bark under your feet. My friend Ed and I used to climb trees together and have long talks about the language needed to describe them: the *treachery* of the yew, the *comfort* of the oak, the extraordinary *flexibility* of the willow. We concluded in the end that the knowledge was more like an emotion, a kinesthetic memory that became part of you.*

For big early primates this kinesthetic memory, relearned in every generation, made a deep impression. Rather than trees preventing us from walking upright, they seem to have been the very reason we do; the climbing frame into which our uprightness evolved.[7] In addition, wood's combination of flexibility and stiffness allowed some primates to build nests. Research shows that they're picky about the species they build a nest in, so first they choose that.† Then they would select a horizontal branch that could support their weight and, lying on the branch, build the nest from the inside. They would reach out to draw thickish branches toward them, break them halfway across in a greenstick fracture, and then hinge them inward, splitting the branches in half and weaving them round in a leafy bower. A short rest, presumably while thinking about the next step and feeling the possible lumps and bumps in the nest so far, is followed by stage two: thinner branches and twigs wound through to strengthen the struc-

---

*Ed, a brilliant musician, knows all about the links between learned movement and emotion. Carpenters and joiners and boatbuilders working with dead trees feel the same.

†Some trees can be antibacterial or lightly psychoactive.

ture and make it comfortable. The result, in around five minutes, is a cup-shaped, safe nest.[8]

Cosseted in these protective, often antimicrobial, potentially psychoactive tree-nests, primates could do something that is incredibly rare in the animal world: sleep. Sperm whales, floating like megaliths, sleep for up to half an hour at a time. Birds sleep with one eye open, employing the opposite half of their brain to watch for any predators. Catnaps are all very well, but they do not allow for deep sleep over long periods. Monkeys, for example, that sleep sitting on branches wake every five minutes or so lest they fall off, but in a nest, both non-rapid eye movement (NREM) and rapid eye movement (REM) sleep are possible for long periods,[*] and the result, broadly speaking, is brain development and dreams.

Humans spend about 20 percent of the night in REM sleep, and it is during this time that we dream. Dreams, for many people, are expendable quirks of the jesting mind, but they have altered the course of human history time and time again. When Xerxes, King of the Persians, was hesitating about invading Greece, we are told he dreamed repeatedly about a great victory; his subsequent defeat paved the way for Greek dominance in the eastern Mediterranean. When the Virgin Mary was trying to escape from Herod she was warned by God in a dream, and Jesus survived his infanthood. And the leader of the Taiping Rebellion, Hong Xiuquan, was told in a dream to rid the world of demon worship, after which he set in motion events that led to the death of 30 million people.

This is just one of the routes by which we make decisions based on a very tree-specific root of being. Dreams are an invaluable way for another world to bolster the waking world. They make a course of action inevitable and are used as an explanatory device for some of the weirder decisions that people make, a super-human intervention. Freud famously interpreted them as a window into the subconscious mind, but recent research suggests that dreams are something more tangible than this, a way of reinforcing your kinesthetic memory,

---

[*] David Samson at the University of Toronto has done many studies comparing types of sleep in monkeys and the great apes. They make for fascinating reading.

the memory that makes you reach for the branch in the right way and swing for the correct distance. Current thinking suggests that the little twitches made by your body in REM allow your brain to map precisely where in the body and the brain that twitch is happening. In other words, in REM we reintegrate our bodies, essential for an ape searching through a three-dimensional tree world for food or bouncing along branches to flee a predator and for an ape developing an idea of self-awareness. Asleep, we develop memories and maps, a shadow world that can impose onto the real one and at times expand it. Trees, therefore, didn't only shape our bodies—they enlarged and shaped our brains every single night we slept in them.

Since we came down from the trees we have been learning how to manage without them—how to replace them in our lives. Sometimes it can feel as though we are now separated from trees, that, although they shaped us once, we have consistently detree'd, and can do without them in our bodies and our lives. Physical tree adaptations like fingerprints and clinging fingers remain, but few of us live now in wooden houses or have diets consisting mainly of fruit. The agricultural revolution replaced fruit and nuts with grains and pulses, the industrial revolution replaced wood fires with coal and oil, and the Bakelite revolution replaced wood with plastic. Scratch the surface, however, and you will find that our domestication by trees is so ingrained in us that society often defines itself around them and that trees still shape us today through the strange handle of our emotions.

The last common ancestor of trees and humans lived about 1.7 billion years ago. Our shared lifestyle ended when one daughter cell engulfed a photosynthetic bacterium and became a light eater, while the other stayed gray and motile, an amorphous blob hunting for food. This provoked a massive and fundamental schism. Even the meaning of life was changed. For the newly green cell biochemical creativity was the meaning of life. It only needed light as an energy source; almost everything else it could make from water, carbon dioxide, and tiny amounts of metal. The non-green cell, by contrast, needed to find chemical energy from elsewhere in order to survive. What we might now call avarice became the defining principle of life. As cells clustered together and multicellularity evolved, this basic chemical difference became increasingly influential; green organisms were structural and sedentary, gray organisms motile and materialistic.

In the last 1.7 billion years since our cells diverged, humans and trees have exchanged little, if any, DNA,* although we still share around 30 percent of our genetic code.† Our shared DNA codes for proteins that maintain or respond to our common biochemical pathways. We share the ability to make and use amino acids such as glutamate (vital for nervous action in humans and communication in trees), the pathways of DNA production, mending, and use, as well as many ion-transport proteins. Trees have continued to develop along the pathway of creativity, producing more and more inventive molecules, and we have continued along the path of avarice, taking more and more of those inventive molecules for our own use. As a result, trees have continued to shape humans chemically in each generation over the last 500,000 years by providing the many compounds we can't make for ourselves. Alongside our structural use of wood, not just our basic food but also the rarer compounds we need to keep ourselves healthy—medicines, supplements, antioxidants, vitamins, and spices—are provided by trees. You are what you eat, and trees have been and remain absolutely fundamental to our diets. They are also, of course, still shaping the air we occupy.

This is more evident in some places than others. After three days and four nights in a hammock groping along the shoal-stiff Rio Negro it felt like heaven to end up in the hands of keen gardeners in São Gabriel da Cachoeira. We wandered round the garden, which

---

* Cross-species DNA exchange is fairly common but mainly occurs through bacteria and viruses. Human skin creates a very different environment than tree bark, and this means that there are few opportunities for horizontal gene transfer, as we are rarely infected by the same pathogens.

† By some assessments. There are many different ways to calculate shared DNA. The idea is complicated by the fact that trees have a massively variable number of chromosomes (from 12 to 138).

was probably little more than an acre, but alongside the vanilla orchid and the chocolate-smelling orchid, the *Catasetum* and *Epidendrum* grew at least ten edible trees, their tastes deliciously expressive of the biochemical magic they had whipped up. There was classic cacao, its yellow pods hanging close to the branch, and also its big cousin cupuaçu (*Theobroma grandiflorum*), a hard, massive brown pod, which tasted of creamy citrus. There was the small coca tree, with white flowers and green glossy leaves, kept as an analgesic, but that had no effect on me when I chewed it. There were the cumari, huge yellow and black, buttery, nutty fruits that felt deeply nourishing. There were the big pods of baraturi, whose pulp tasted like parma ham and melon mixed together. There were jacar and jaubo; maracushar and castane da picia; there was abacacharana, a tree yellow outside and pink inside that smelled of Jerusalem artichokes; there was deep purple açai, coffee and paciumba, coriori and jawakan, and much more besides. Many of these fruits had evolved to feed up megafauna, but that didn't matter; now they were fattening me. Ninety percent of these trees could be found in the tangled forest within one mile, and in the center of the town two huge Brazil nut trees dropped cannonball fruits, gigantic trees, 400 years old, that had been planted by the first Jesuit missionaries from Europe.

A couple of days later I went out with Giovanna, who had made the garden, to find some very rare trees that only grow along the white sand of the Serra do Curicuriari nearby. We walked for a couple of hours to get there, and along the way Giovanna chose seeds and saplings to take back for the garden. The shaping of trees into a garden seems to flip the idea of trees shaping humans, but does it? The botanist Michael Pollan famously underwent a gestalt shift when, on his hands and knees weeding potatoes, he realized that the potatoes had domesticated *him*. Yes, in the end he would eat some of them, but in the meantime he was on his hands and knees day after day, weeding, fertilizing, untangling, and protecting the potatoes, making his garden into a potato utopia. By the human, for the potatoes.[9]

In the Amazon, the idea of humans being domesticated by certain species of trees is everywhere. Brazil nut trees find it hard to compete in dense forest without human intervention. Just as the trees literally shaped the size of the huge bees that pollinate their creamy flowers, so their vastness and charisma, the selenium and concentrated oil of their sapid nuts meant that humans planted and nursed and cleared for them. There is no "virgin" forest in Central America and most of the Amazon Basin: archaeologists are beginning to show that every single part had been shaped by humans long before Columbus landed in the Americas. The earth is often the *terra preta*, enriched with carbon and phosphate by people over thousands of years to nourish trees, and many islands in the forests near areas where the Kayapo Indians live are in fact man-made. *Apêtês*, as they are called, start as compost heaps. In due course a mound of earth 7 feet in diameter is formed, which is then planted with useful trees. More trees are planted around the edge of the island, until a patch of forest containing all the necessities of human life is created. Fruit trees are planted to attract game, and the general tendency is toward an increased diversity of trees.*

Much of the chemical manipulation that trees practice on animals remains appropriate. The striking relationship between the cacao tree and monkeys, for example, appears to have been seamlessly transferred into humans, even though our genetic paths parted from the monkeys of South America 5 million years ago. The debate still rages over the exact dates when humans first traveled to the continent of America, but new evidence suggests that they, like the cacao tree, were sheltering in small warmer patches of South America 20,000 years ago.[10] The cacao tree was the keystone species of many pre-Columbian South American cultures. Could it be that as the world warmed up cacao and humans expanded their range together? Was it simply that the seeds of that very early interdependence survived in our DNA, encoding a human protein (an adenosine receptor) that was chemically reshaped by the rings and oxygens of theobromine? Or was it the re-forming of an old relationship of compatibility and

---

* Personal communication from Ghillean Prance, former director of Royal Botanic Gardens, Kew, London, and someone who has an extreme respect for the ways in which the myths and beliefs of Indian groups protect and conserve the forest.

attraction: our eyes drawn to the red and yellow of the pods, their convenient positioning by the tree close to the trunk an invitation for us, like our monkey ancestors, to pick them and spread the seeds?

Certainly, the allure of cacao is hardwired into our nervous systems. At whatever stage the primate responsibility for dispersing *Theobroma cacao* transferred from monkeys to humans, it was firmly embedded by the time most pre-Columbian cultures developed. It was fundamental in Aztec society, where the tree was associated with the god Quetzalcoatl, who, much like Prometheus and fire in Greek myths, was damned by the other gods for sharing chocolate with humans. Meanwhile, the Mayan kings were chocolate connoisseurs; their beautiful cacao drinking bowls had glyphs that lingered lovingly on the properties of the cacao: foaming cacao, tree-fresh cacao. Europeans were equally susceptible, and the majority of cacao trees are now found growing in West Africa, part of a huge global trade.* Cacao trees currently grow everywhere in the world they can survive, from southeastern Mexico to the Philippines, the Ivory Coast, Angola, and India, and we plant them, nurse them, and consistently destroy their competition and pathogens.

How many more examples of trees shaping human behavior, domesticating or taming us, are there? The soothing scent of a new book is a response to the smell of lignin in the paper, specifically a subunit, vanillin, that is also released into some whiskies aged in oak and concentrated by the vanilla orchid, which grows on trees.[11] Unlike most orchids, which have seed dispersed by the wind, both bees and animals disperse vanilla orchid seed—one reason for its extremely wide distribution. This means that the smell of both cupcakes and books can be traced back to the appeal of our woody nests. The distribution of the valonia oak, from the Zagros Mountains in Iraq to the tip of Italy, can be traced back to the transhumance pastoralists who traveled along the mountain ridges forming spines through Anatolia, past northern Greece, and down through Italy. Cacao was a

---

* Trees have lost out to wheat, barley, and rice as key suppliers of our food chain, but have they really lost out? Their esoteric but essential contributions to our diet—less staple, more indispensable extra—mean that they are widely distributed.

little bit lucky in its chemical choices, but by being portable it gained the upper hand, and so subtly that humans aren't even complaining. There are countless examples from around the world of trees becoming the keystone species of societies: the pistachios of Gaziantep in Turkey, the olives of Crete, and the walnuts of Kyrgyzstan have all caused whole human communities to organize their lives and culture around them.

In the same way, I have a friend who has been domesticated by apples. Unlike most apple growers, who stick to clones that are known to be delicious, he grows endless pippin apples from seed, nursing vast numbers of small, hard, bitter apples in search of the precise genetic combination that will produce the perfect apple. He has devoted his life to this quest, a glorious example of human free will in the service of trees. Apples are a major source of one of the essential chemicals we must take in through our diet: vitamin C. The chemical pathway to get to vitamin C, a relatively simple molecule, takes an estimated twenty steps. To construct the proteins to take those steps takes an estimated 2,000 reactions. To construct the proteins to construct the pathway to construct the chemicals takes an estimated 250,000 base pairs, the basic code in DNA,[12] and all this must be powered by the captured light of the sun.* Humans could never spare the energy to synthesize this for themselves, but for trees it is worth it, first as protection against free radicals generated by the sun, next as protection against small predators, and finally through its demonstrated power to domesticate large animals, including humans.

The apple tree originated in Central Asia, where its wild ancestor still grows. During the last glacial maximum about 23,000 years ago, *Malus sieversii* survived in an area around the Ili River, which flows out of the Tien Shan to Almaty in Kazakhstan. After the glacial maximum passed it expanded its range, spread by bears and horses, but most of all by humans. It comes from a forest near Almaty that,

* This pathway is named by biochemists the Smirnoff-Wheeler pathway; a joke about needing lots of vodka not to cycle off in despair just looking at a representation of it is a staple of the biochemistry lecture theater. The word *pathway* is a human misnomer because it implies a route from A to B, whereas this route may branch or cut itself up at any stage, replacing itself, inverting itself, even running backward.

like the Amazon but on the other side of the world and in a very different environment, abounds with the wild ancestors of the fruit and nuts we know today: plums and almonds, pistachios and apricots. As you wander up this river you can see the feral hints of all the different apples you might see in a supermarket: the pink-red of Pink Lady or the green-white of a Granny Smith, a hint of yellow or red, even the brown skin of a russet like my favorite, Ashmead's Kernel. Tasting an apple from this area gives you vistas of the ways that apples could have developed. The river is an avenue of potential tastes—rose-like, sour-pear, aromatic, bletted sweetness.

If the apple had stayed in the Ili valley it might still be unknown to us, but instead it was cultivated in the Tien Shan for around 5,000 years, during which time genes from the wild crab apple crept in. Certain apples were sweeter than others, so humans started to graft branches from the sweet apples onto other robust trees. They were grown, as we know from apple cores carbon-dated to the tenth century BC, in orchards in Iran and Iraq and Israel, prized by the writers of the Old Testament, mythologized in the Norse countries, and burnished by the poets of Ancient Greece, who made the golden apples of the Hesperides, brought by Gaia to the marriage of Zeus and Hera as a wedding gift, the most prized of all gifts of the earth. In twelfth-century England apples were the fruit trees of the Isle of Avalon, mythical home of King Arthur, although by 1518 the literary associations of apples were less elevated; in one of the first pamphlets printed in England, Wynkyn de Worde's "The Crafte of Graffynge and Plantynge of Trees," the author advises those who want their apples to be sweet to "sponge the roots with pigge's donge"—possibly the first reference to the use of a specific high-phosphate fertilizer.

The same tree spread into China, where it is also very popular; crossed with another variety of *Malus* it is grown there for its softer, sweeter fruit. And when the apple crossed to South America in the sixteenth century it flourished in the Chiloé Archipelago, while in the United States it arrived in 1640 in Boston, traveling with the Reverend William Blaxton. It picked up a proselytizing poster boy in the form of Johnny Appleseed, an eighteenth-century orchardist who canoed down the Ohio River with apple saplings and established nurseries as he went. No tree could ask for a better seed disperser: he introduced apple pippins to Pennsylvania, Indiana, Illinois, West Virginia, and Ontario, and with him he carried the apple myths, the symbolism of the tree present in the Garden of Eden, a mystical conversion of disobedience into New Life. Johnny Appleseed was clearly remarkable, but his trees were probably not particularly tasty—it was as an allegorical figure of abundance in the absence of materiality that he made his mark. And although Frank Matthews, the UK's great apple nursery, are passionate grafters—and commercial apples are nowadays grown on trees at waist height for better picking and made as rectangular as possible—most of the apples people will have nursed across the 10,000 years of our mutual domestication will have experienced human generosity, a particular form of social egalitarianism we would do well to imitate.

Orchards and fruit trees are obviously useful to humans, but why is it that we also recognize trees as metaphysical objects—pathways between the physical and the ideological? At about the same time as Johnny Appleseed was canoeing down the Ohio, trees were shaping human society not just literally, but as powerful symbols of liberty. "Spent the Evening with the Sons of Liberty, at their own Apartment in Hanover Square, near the Tree of Liberty," wrote John Adams on 14 January 1766.[13] Adams was a leader of the American Revolution and would later become the second president of the US. In August the preceding year, a crowd had gathered under a large elm tree at the corner of Essex Street and Orange Street (now renamed Washington Street) in Boston and had hung an effigy from the tree labelled "A.O." for Andrew Oliver, the unfortunate colonist chosen by George III to impose the Stamp Act, a tax on every document printed. The tree was planted in 1646, just sixteen years after Boston's founding. Everyone traveling to and from the city by land would have passed it, as it stood along the only road out of town, Orange Street. (Boston sat on a narrow peninsula until the 1800s, when the Back Bay was filled in.) Though no measurements of the tree survive, one Bostonian described it as "a stately elm . . . whose lofty branches seem'd to touch the skies," and it became a potent symbol of revolution. When news of the Stamp Act's repeal reached Boston in March the following year, crowds gathered at the liberty tree to celebrate. The bell of a church close to the tree rang, and Bostonians hung flags and streamers from the tree. As evening came, they fastened lanterns to its branches: 45 the first night, 108 the next night, then as many as the tree's branches could hold. It was chopped down in 1775 by the aptly named loyalist Nathaniel Coffin.

However, other liberty trees had been planted across America, and (perhaps echoing this) in 1789 a formal campaign of liberty tree planting was started in France as a rallying point for the revolution. Called Napoleon trees, they were poplars or plane trees, oaks or elms, all planted in the center of towns, and the pattern they created across France was itself mapped as a tree. "The tree of liberty must be revived from time to time by the blood of patriots and tyrants," wrote Thomas Jefferson,[14] and his words were invoked more literally at King Louis XVI's trial, which led to the guillotine.[15]

Unfortunately, many of the trees died after they were planted, and an aquatint by Jean-Baptiste Lesueur shows us why. The pictured tree is planted, with pomp and music and an excellent party, in full leaf in a small patch of soil among lifted cobblestones, in a hole far too small for a twenty-year-old tree. As a result the National Convention ordered, by a decree of 22 January 1794, "in all the communes of the Republic where the tree of liberty has perished, another will be planted by the 1st of Germinal." And that after the planting the citizens of each commune would be responsible for watering and protecting the tree, "so that the tree would flourish under the aegis of French liberty."

Liberty trees were planted as symbols of revolution and change, but also continuity. A few years ago I cowrote a paper called "The creation, content and use of urban tree strategies by English local governments." It was a dry piece of work, but my part in it was leavened by interviewing twenty-five tree officers from around England, from Solihull to Sheffield, Kensington to Cornwall, the majority of whom were genuinely amusing people. Tree officers are the local government employees responsible for planting or removing all trees within the borough, so a sense of humor is essential, and tree strategies are a way of ensuring that councils plan for the multigenerational time-scale of trees. A tree planted in a city can need a lifetime's care if it is to survive.

Compared with the speed of change of technology or architec-ture, trees change very little. Perhaps some of our very rootedness as humans, our wish to temper the excitement of technological ad-vances with a grounding in nature by planting trees around our new-est cities, is part of how trees have shaped us. It's linked to our very real and serious fear of climate change and the stability of our habitat breaking down. We wish to retain a rooted part of our environment; unable to migrate away from danger, trees can act as the canaries in the mine. Certainly, there are trees that have seen the rise and fall of cities, and clones that have seen the rise and fall of empires. One beautiful example is an elm tree in New York City that is older than the city itself. It occupies a patch of real estate worth millions, a mile away from World Trade One in Washington Square Park, the heart of the city, but because it is such a beautiful tree it has been left in place as buildings have risen around it. From a broad trunk its two strong leaders spread out into a wide canopy, and when I was in New York in April it was just coming into leaf, a light dusting of spring green against the red brick of the square.

In 1990 the New York tree was determined to be 310 years old, which means it has probably occupied its current site since 1680, sixteen years after the British conquered New Netherland and changed its name to New York. The English elm isn't native to the United States or to the UK either, but the elm is likely to have been brought from Britain not as a seed but as a cutting from an elm on an English farm. Elms are good at reproducing asexually, and most English elms have identical DNA; they are just one clone called *Ulmus minor* 'Atinia', for which we luckily have a written record covering almost 2,000 years.

This clone arose in Spain, probably just a seedling that grew up by chance. It was spotted by someone who brought it to Italy, where its introduction was recorded by the Roman farmer and writer Colu-mella in AD 60, as he sat and wrote at his farm at Ardea, 20 miles outside Rome. Elm was popular as a structure for vines, for hedges, because it suckered along to make a natural row, for tree hay to feed cattle, and also for drains, as the wood is very resistant to rot.* The Romans brought the same clone to Britain, where it was planted extensively, and after the Enclosures Acts, from 1604 onward, it became the sign of a boundary.

The first British settlers in New York probably planted it on their farm to create a hedge tree in the way they would have been familiar with from England. Chemically, it must have endured and adapted to so much: an English farm, perhaps salt and dark on the transatlantic crossing, the microflora, secondary metabolites and climate of a newly cleared piece of American forest, and finally, in about 1880, the extraordinary increase in heat and decrease in light that came with the construction of Washington Square. Minetta Creek, which would have been close enough to affect the pressure of water reaching the tree's roots, was built over, and the Minetta Tavern, favorite watering hole of Ezra Pound, sprang up. Traffic, increasing smoke, and all the particulate matter of a big city drowned the tree and the insects it interacted with; the land it was on went from valuable to priceless—and yet remarkably it survived.

The tree was constrained by various factors in the early days of its growth, not least the lack of a specific inherited bioflora to associate with. However, this isolation paid off. Alone and cut off from other trees, it was not infected by the Dutch elm disease that destroyed almost all the old elms of the UK. Like the United States, it became larger and more successful than its forebears. And so it is that something that is likely to be genetically identical to one of the foundational trees of the Roman empire—foundational in drain-

---

* Even in the worst and wettest or most anaerobic conditions, elm does not rot because it has firm compartmentalization throughout its trunk. Unlike oak or other hard woods, elm gets more resistant to deshaping—not less—when the pores of the wood are clogged. And the ability of trees to work as portable conduits of material was important. Other early materials varied massively from place to place. A clone of the tree that Columella found so helpful on his farm could be planted in England and grow in almost exactly the same way as its predecessors—it shared the same DNA.

ing the marshes and saving Rome from malaria, foundational to the Roman myth of the homestead, and the bucolic idyll where Horace and Martial escaped the rat-race—was also able to grow to fruition at the heart of a great new world city.

Like orangutans returning to one tree over and over again, a tree can be central to societal custom, while a tree's peculiar and often ill-defined status—never truly owned but often existentially necessary to certain people—made it an interesting point of contention in revolutionary times. Most often the line fell between the sustainable use of a tree and its non-sustainable removal. In countries as diverse as the UK, Iran, and Japan the right to the wood (the regrowable part), as opposed to the timber (the non-regrowable trunk), was governed for centuries by tight custom, only to be crushed by interlopers writing laws. In the UK after the Norman conquests, cutting living wood carried the death penalty (although, as Oliver Rackham, the great ecological historian, points out, this was rarely if ever enforced). The woods attached to Zoroastrian fire temples were lost with the advent of Islam, and the Japanese novelist Yasunari Kawabata writes about the subtle collapse of custom in Japanese society when carefully shaped wood is no longer prized. All these were examples of tree care and benefit from the tree being uncoupled. Even Karl Marx wrote in the introduction to *Das Kapital* that it was trees that had led him to think seriously about property.*

Marx's articles on wood theft for the *Rheinische Zeitung* critiqued the Rhenish forestry code for privileging one form of entitlement, the law of possession, over another, the customary right to usage.[16] The "small, wooden, spiritless and selfish form of interest" had triumphed over the hybrid forms of possession that had allowed people of various different classes to benefit from the same common resources. In the case of forestry in France this meant 37,328 verdicts

---

* I'm indebted to Jessica Hao for this passage. As noted in Christopher Clark, *Revolutionary Spring: Fighting for a New World, 1848–1849*.

in cases of wood theft between 1824 and 1829 and more than 14,000 in "other forest-related offences." All over France the *Code forestier* of 1827 blocked people from collecting firewood, pasturing animals, extracting timber for building materials, and lighting fires within 650 feet of forest borders. In many places this led to revolution. In the Pyrenees the Guerre des Demoiselles, or War of the Girls, limped on until 1872, as men dressed as women attacked the forest guards in a low-level guerrilla war in protest against the new law.[17] They put on white-paper masks, daubed their faces with thick paint, and untucked their long white shirts and bound them at the waist with sashes to echo the traditional forest spirits known as *demoiselles* or *dames blanches*. The prohibition on managed fires, which were normally set at the end of the winter or in early spring, also caused conflagrations, the most notable in 1891, when 150 acres of forest caught fire.

The transformation of a dispute about picking up wood into an attack on the principle of bourgeois ownership is Marx's genius, but the difference between protection of wood and protection of timber is a profound one. Trees, unlike animals, can recover from a lot, and coppicing, pollarding and shredding, as well as picking up wood off the ground, all allow for a non-depleted resource. Removal of timber, on the other hand, can lead to deforestation, soil erosion, and the utter destruction of an ecosystem and the culture that relies on it. In Lombardy in 1848 the clearing of woods as a result of economic pressure led to a chronic shortage of firewood and a smorgasbord of natural disasters.

Two years ago I saw something similar happening in eastern Turkey, near Siirt, where mountainsides of shredded trees, the smaller branches cut off year after year for firewood to leave the tree still thriving, had been clear-cut, perhaps simply as a result of the introduction of a chainsaw to the area. The results had not yet started to show, but if comparison with other areas holds true, there will be landslides and loss of topsoil and subsequent economic collapse within the next two years, and the landscape will be so completely changed that it will be as though all those hundreds of hundred-year-old trees never existed. When our need of trees is ignored, our suffering matches or exceeds theirs.

In my local Anglican church of Bishop's Nympton in Devon is a small wooden statue of a saint, which sits above the door of the vestry in a cobwebby niche. Carved out of oak, it is supposedly Saint James, brought back from Spain by a roving vicar in the thirteenth century. He is clearly old, signs of wear deeply ingrained, and holding something indistinguishable—a hat, perhaps, or a book. Every time I see the statue it reminds me of the young mountains of the Hindu Kush, where an animist community, the Kalash, lives and protects cedars of antiquity and outstanding beauty. When Alexander the Great's soldiers, who had come over the mountains from Afghanistan and were freezing to death, found cedarwood planks on the ground, they burned them, whereupon they were attacked by the Kalash; the cedarwood was the residue of coffins, in which the Kalash expose their dead on the mountainside. Alexander and the Kalash came to an agreement, and Alexander moved on to conquer King Poros. Since then the Kalash have survived as a small community, sometimes embattled and even attacked, occasionally converted, but mostly remaining animists, who believe that trees, animals, and the natural world have souls. Their coffins are still made of cedarwood, beautifully carved and left open to the air, and occasionally their shamans will carve wooden statues of people out of cedar trees. In spite of this, the cedars in the Kalash Valleys, the prized deodar or Indian cedar, are some of the most beautiful and best protected I have ever seen.

Animist communities like the Kalash are often small and remote, but they are not underdeveloped. Their understanding of what it means to protect a tree does not mean planting millions or never cutting one down. It means an aesthetic appreciation of what a tree is, and acute observation of how trees function. Trees are not anthropomorphized by animist societies, as they sadly are by so many people who have grown up not looking at them. They know that our nervous systems are not the same and acknowledge that trees' abilities are of far more importance to the environment than ours.

Trees have profound agency over earth, wind, fire, and water in ways we can only partly elucidate. Scientifically trees are full of surprises. To try to apply physical laws to them is to invite disaster—there are so many species with their own peculiarities, and each has its own laws of behavior. Defining them chemically is equally risky—we will never catch up with the new compounds they invent while we sleep. Classifying them genetically is, as any *Sorbus* enthusiast will tell you, a slippery business. An occasional glimpse of root in a cave roof, the taste of an unusual fruit, or a mirage-like cluster of palm trees in the desert gives us a glimpse of the living unknown; a life-form that shaped the world we grew into, that wove it around us into a peaceful form, but is still largely incomprehensible.

And yet, when we reach out to grasp a branch it fits into our hand. Whether vast or small, we can understand the shape of trees, and we can see our world reflected in their differences. Most of all we can see glimpses of them in us: a life-form that shaped us whoever and wherever we are, in such a way that we are entirely different from them. They help root us and make sense of our coming and going, our beginnings and endings, our birth and death.

Nympton means sacred wood, and I like to think that some of the old paganism seeped through. It doesn't have be overt: a fundamental bond with something can run very deep. I also like to think that the aesthetics by which we render ourselves sensitive to a tree can draw on the product of scientific inquiry, that we do not necessarily have to undo the work of the past to look for a road into the future. I hope that the culture of trees runs so deep in all humans that the most egregious gougings—the madness of cities or the absorption of tech—cannot tear it out.

In 1890, when he was deeply depressed, Vincent van Gogh wrote to his brother about one of his paintings, *Tree Roots*. He explained that he had chosen his subject because he wanted to express something of life's struggle. "My life too," he wrote, "is attacked at the very roots."[18] He had settled in Auvers-sur-Oise, a village an hour north of Paris popular with painters, where the thatched cottages of the old French world mingled with the new middle-class houses. It was not, you would think, the ideal spot for animism, and his preoccupations were not of nature; he was saddened by people, who ignored his work, by poverty, and by his brother's increasing ill health as a result of syphilis. He was saddened by the painter Gauguin, whose long-awaited arrival was a terrible letdown—Gauguin barely glanced at his paintings and told him to use less green. Nonetheless, on the day of his suicide, 27 July 1890, he painted a picture from life in the Rue Daubigny, above the manor of Colombières, where the road cut through a wood of larches to leave a steep bank with tree roots exposed. The painting is his last masterpiece, gnarled bright-blue tree roots and green foliage bursting out of a taupe bank. It is a picture of trees holding it all together, making sense out of a bewildering world.

Epilogue

I was supposed to plant trees, and I didn't.

Primo Levi, *Unfinished Business*

In 2019 a friend took me to Sidakan, a village high up in the mountains near the Iranian-Iraqi border, and introduced me to his father, who gave me a huge fat acorn that I put in my chest pocket and promptly forgot about. The acorn traveled with me to Baghdad, where I realized that I shouldn't take it out of the country and at the airport I hastily pressed it into the hand of a man traveling to Najaf. It was a pang, and I pined after the acorn and the beautiful tree it would have been. A few years later I was overjoyed when an acorn of the same species arrived in the post from an aboretum in the UK. It was equally as big and beautiful, but it seemed dried out, the shell slightly cracked and battered from its journey, so I didn't have much hope as I hung it up in a plastic bag in damp sand.

Three weeks later I poked around in the sand and found it; to my joy the acorn had expanded and a wormlike, pale green tap root was emerging tentatively from one end. I planted it in a clay chimney pot on top of a pot so that the root could go down deep, filled them both up with sand and limestone gravel and some leaf mold from the oak woodland in the valley—in case it could find some useful ectomycorrhiza among these relicts of distant cousins—and watched anxiously. The enormous acorn produced hairy silver-pink leaves with serrated edges every April and dropped them in November. Each year it seemed like a triumph when it came into leaf, and it grew achingly slowly, an inch a year. When it made its first proper rough brown bark it seemed like a watershed moment. When it took a kink to the left I feared for its future. Finally, this year my father and I dug a big square hole, filled it with sand, more oak leaf mold, some limey shale, and topsoil and planted the contents of the pot. It had a root over 3 feet long, and a nascent trunk barely 2 feet tall. It is now unpleating perfectly formed French-green leaves and reaching up to the light.

This tree is linked by kinship and nature to trees around the world. It is part of the weft of life, a lignin lattice, connected to Iraqi oak trees that share its DNA and English oak trees that share its symbiotic fungi. It is already fixing down infinitesimal amounts of carbon dioxide and altering the atmosphere. Over hundreds of years it will reach out and down, its tap root worming into the bedrock to split the neat rock, finding caches of minerals like hidden gold and forming a downward passage for water. Meanwhile its shallower roots will be reaching out to the trees around it and forming the apex of a pyramid of interactions with other plants and fungi in the soil. If an insect escapes my eagle eye and bites into those tender leaves, a chemical signal will be released into the air, detected, and probably half-understood by the trees around it. Overhead an imperceptible storm cloud is already gathering: water both released and gathered by the tree to fuel its growth up and out and down over hundreds of years. If I let it.

But there's no question about that: I'm all in.

Notes

Introduction

Epigraph: Hildegard of Bingen, quoted in Newman, *Voice of the Living Light* (1998).

Chapter 1: Trees shaping water

Epigraph: Nāzik al-Malā'ikah, "At the End of the Stairs" from *Shrapnel and Ash* (1949), in *Revolt Against the Sun: The Selected Poetry of Nāzik al-Malā'ikah, A Bilingual Reader* (2020), p. 40.

1 Photosynthesis is doing something very hard, so in detail it is an inescapably complicated business. The best simple explanation is in Nick Lane's *Life Ascending* (2010), while for more detail Nicholls and Ferguson's *Bioenergetics 4* (2013) is unbeatable.
2 Beautifully explained in Boyce et al. (2017). It highlights how easy it is to fall into a trap of projecting modern climatic conditions and tree types back into an earlier, very different time.
3 Gymnosperms, in spite of their rarity, still act as keystone species around the world. *Araucaria angustifolia*, the Brazilian pine, is a critically endangered gymnosperm that is native to the south of Brazil and forms the emergent trees of cloud forests like those of the Canaries in which the araucaria acts as a nurse plant and a fountain tree, allowing laurels such as *Ocotea catharinensis* in their turn to grow and capture the water blowing in with the trade winds off the Atlantic.
4 Some gymnosperms, such as *Gnetum urens*, the bell's bird-heart vine, adapted fast to fit into the new rainforests of South America and became climbing vines. And while we can broadly think of gymnosperms as stiffer and less yielding, it's unsafe to generalize. In one mysterious example, the swamp cypress grows "knees," great angular root cones,

only when its roots are wet and dry by turns, but the purpose is obscure, as the knees apparently offer neither support nor air. For more on this enigma see Briand (2020).

5 The peak is named after Gara, Princess of Gomera, and Jonay, Prince of Tenerife: star-crossed lovers of Guanche folklore. Their marriage was forbidden by the omen of fire from the volcano on Tenerife, but Jonay swam across the sea between the islands to be with his beloved. Hunted by their fathers, they were driven to the top of the peak where they killed themselves. Related in Ruiz (2017).

6 This paper shows the pictures in fine detail: Klimko et al. (2008).

7 Waxes play an essential role in the plant world because they can minutely adjust permeability and transpiration rates. For more on their role in regulation of dragon's blood leaves, see Jura-Morawiec et al. (2020), and for a more general discussion of wax plugs as a tool for evolutionary spread, see Brodribb et al. (1997).

8 "*Está una nubecita siempre y sobre un árbol. Cuando está junta junto con el árbol, parece está algo alto del árbol, cuando se desvían parece estar junto del y casi todo lleno de niebla. / . . . / Aquella nublecita hace sudar y gotear todas las hojas y ramas del árbol, todas la noche y el día, más a las mañanas y a las tardes, algo menos a mediodía, cuando se alza el sol . . .*" From Las Casas's *Historia natural y moral de las Indias* (sixteenth century) quoted in Acosta Baladón (1973), p. 97.

9 Kolby Jardine, working at the intersection of biochemistry and atmospheric science, perched for days on end on the Amazon Tall Tower Observatory in Manaus, Brazil, to measure these different isoprenoid interactions; see Alves et al. (2016) and Robin et al. (2024).

10 Data recorded in Shilling et al. (2016). For more analysis see Shrivastava et al. (2019).

11 Gollut in 1592 blamed deforestation in the Dôle area of France (iron-founders consuming trees faster than they could grow) for increased rainfall in the previous twenty-six years. Braudel (1966) quoted in Grove and Rackham (2003), p.11. Similarly, deforestation is often believed to cause floods.

12 Quoted from the original paper, Makarieva and Gorshkov (2006).

13 See Flores et al. (2024) and Smith et al. (2020) for the latest empirical evidence.

14 A brilliant and highly readable account of the Devonian Old Red Sandstone of Ireland can be found in Sleeman et al. (2004), Chapter 7.

15 As Bill Stein's seminal paper in *Current Biology* poetically puts it: "several well-articulated fish have been recovered near the largest trees, seemingly impounded by them." Stein et al. (2020).

16 "We are impeding our passage down the smooth stream by catching at some branch or root," wrote Virginia Woolf in *Mrs Dalloway* (1925). In those days rivers flowed unhindered by trees, and they ran in different ways, totally different shapes.

17 It was shown in 2016 that this apparently impossible feat was possible because of the ability of water to withstand a vacuum. See Boatwright et al. (2015).

18 For more on the ability of plants to hear see Schlanger, *The Light Eaters* (2024).

19 Brilliantly illustrated by Vincke et al. (2008) in a stand of Scot's pine using the water table fluctuation method.

20 For eucalyptus and their eye-opening effect on the water table see inter alia Hoogar et al. (2019).

21 de Witte et al. (2012).

22 The result is that they produce 2 ounces per square foot squared per acre—higher than the lower level of temperate forest production and exceptionally high for a desert. Normal sequestration of carbon at the equator is exponentially larger than at the poles as a result of species diversity. See Buras et al. (2013).

23 Decombeix et al. (2011) describe the fossil Antarctic with some aplomb.

24 The gymnosperm database is currently an excellent and up-to-date source: https://www.conifers.org/topics/vegetative.php

25 A very intense debate surrounds the mechanisms that allow roots to obtain sufficient air and the role that this had in their development. Unlike with water, air has no dedicated air movement channels, and so most of the time the gases needed for cellular functions, like oxygen for respiration, must be supplied either by diffusion from the outside of the tree or by the movement of gas dissolved in water. Boyce et al. (2017).

26 Even so, they often become affected by high levels of salt, which sucks water out of the plant by osmosis. In an Indian mangrove, *Avicennia officinalis*, 95 percent of salt in the water taken up by the plant is excluded and moved to the cortex, but any that remains and does accumulate in

the shoot is then concentrated in old leaves that the plant then sheds, sometimes to the extent that salt crystals can be seen on these old leaves. See Tomlinson's *Botany of Mangroves* (2016).

Chapter 2: Trees shaping soil

Epigraph: Intizar Husain, *Basti* (2012).

1 For more on the many weird and wonderful ways that the earliest trees grew, see Normile (2017).
2 Boyce et al. (2017).
3 In this line from "The Waste Land" (1922), T. S. Eliot evokes the most fundamental sign of death that he can. His images of *the dead tree gives no shelter* and his questions *What are the roots that clutch, what branches grow Out of this stony rubbish?* are taken as an allegory of the inversion of life after the war, but can also be taken more literally.
4 The red sandstone of Svalbard acts as a near continuous record of sediment from the Silurian to the Mid-Devonian and is one of the clearest illustrations of how trees started to change the landscape at the start of their development. Davies et al. (2021).
5 For more morphological changes that have inspired the differences that we see and experience, see De La Torre et al. (2019).
6 Adolpho Ducke (1935): 331, quoted in Cardoso et al. (2015). All identification in this passage relies on this excellent paper.
7 By contrast conifer leaves are often very long lived, with the Chilean *Araucaria araucana* holding the record, with leaf life spans of twenty-five years (Lusk, 2001). Phosphorus is the major limiting factor for growth of the entire Amazon rainforest, which, as the Amazon is a turbulent, totally heterogenous, green blanket, covering an area of 2.6 million square miles (nineteen times the size of Germany) is a big claim. However, in 2022 some thirty-one authors, headed up by Hellen Fernanda Viana Cunha, a pioneer of genius in thinking about the Amazon as a whole, published definitive proof in *Nature* magazine that the net primary productivity of the rainforest increased exclusively with the addition of phosphorus to the soil. Fine root growth increased by 29 percent,

canopy productivity by 19 percent. In the paper the authors noted that the CO_2 fertilization effect—the faster growth of trees as CO_2 levels increase—is prevented by this lack of phosphorus. In other words, if the Amazon were supplied with more phosphorus, it might well grow fast enough to soak up enough CO_2 to counteract some of the effects of global warming. Cunha et al. (2022).

8 Huaráca Huasco et al. (2021).

9 Alfred Russel Wallace's illustrated tome *Palm Trees of the Amazon and Their Uses* (1853) contains many eccentric stories. Both rival and collaborator with Darwin, he went on to posit ideas about natural selection, mainly shaped by his observations in South America.

10 Some of this research was funded by the United States, parts of which are directly affected in changes in weather patterns caused by deforestation. Excellent papers on the subject include Freitas et al. (2005) and Houlton et al. (2018).

11 Rocks such as mica-schist, derived from marine sediment, provide almost a quarter of nitrogen in terrestrial ecosystems. In 2018, scientists readjusted the textbooks to show that, but that still leaves 75 percent of organic nitrogen to be found from other sources reflecting this new discovery. The study suggested that more than a quarter of global nitrogen used by plants comes from the earth's bedrock. Ahlgren. (1994).

12 Chen (2021). See Kniep (2007).

13 But whereas hemoglobin works in humans to supply oxygen and keep our cells alive, the alder trees growing along the river were binding oxygen to protect their most valuable ally—the nitrogen-fixing bacteria they house in their roots. For a long time the scientific community debated who exactly—bacterium or tree—was supplying the hemoglobin, but finally the results piled up. Santana et al. (1998).

Chapter 3: Trees shaping fire

1 For Rackham quote, see Grove and Rackham (2003), p. 54; *Abies equitrojani* is properly classified as *Abies nordmanniana* subsp. *equi-trojani*. Farjon, A. (2001).

2 du Châtelet (1744); I had overlooked Isabelle Bour's excellent translation,

for which I am now very thankful. "She came very close to describing fire in much the same terms as our modern concept of 'energy,' an entity animating all substances." Emilie du Châtelet, *Selected Philosophical and Scientific Writings* (2009). Her biography can be found at Detlefsen, "Émilie du Châtelet," *The Stanford Encyclopedia of Philosophy* (Winter 2018 edition).

3 The chemical reaction that causes glow-worms to give off light has been beautifully characterized, and luciferase, the enzyme involved, was the subject of a seminal biochemical paper by Elena Conti when it was crystallized in 1996. The enzyme oxidizes a compound, luciferin, and produces all light and almost no heat.

4 Smart et al. (2002).

5 Read et al. (2021).

6 See Satyal et al. (2013) for an excellent analysis of volatile constituents of *Pinus roxburghii*.

7 Various handy sites show the breakdown mix of volatiles including http://sitem.herts.ac.uk/aeru/ppdb/en/Reports/2012.htm. Whiteside et al. (2015).

8 In George Stewart's book *Fire* (1948), about a forest fire, he describes lightning building up in a cloud: "the cloud was fuming and raging inwardly much like a stupendous steam-engine, coupled with a monstrous dynamo, neither provided with any continuous outlet for the energy which they piled up. Every moment myriads of rain-drops formed and fell through the cloud, only to break into smaller droplets and be whirled upward in swirling vortexes of air. By these internal forces a positive charge of electricity built up in the highest levels of the cloud and a negative charge in the lower levels, particularly at the forward edge.

"Meanwhile, as action demands reaction, the negative charge attracted an equal positive charge to the earth beneath the cloud. Like an invisible shadow, this charge drifted along beneath, up and down the canyon-sides, across the ridges." Stewart (1948), p. 9.

9 For more on the petrified forests of Arizona, see https://www.nps.gov/pefo/index.htm

10 See Balter (2016).

11 Georgia O'Keeffe in her own words, "My faraway nearby." *Tate Etc.*, issue 37.

12 Lyons et al. (2009) map the different biogeochemistry of resins from various tree populations, preserved for many years.

13 New Zealand's past papers can be accessed at https://paperspast.natlib
.govt.nz/newspapers/EP18960211.2.13

14 For the fascinating story of how amber deposits round the world can
show us the shimmering, moving gold arcs of the gymnosperms see
Declòs et al. (2023).

15 More comprehensive discussion of amber and what it can tell us about
the Carboniferous can be seen from Seyfullah et al. (2018).

16 Named by the botanist Philip Miller in 1768 after Aleppo, queen of
Syrian cities. Philip Miller probably got the seed from the consul at
Aleppo; he mentions that many of the Aleppo pines planted in England
died in the great frost of 1740, and the greatest that survived were in the
warm microclimate on the south side of the Sussex Downs at Goodwood. (The two largest I have seen are in Goodwood.) The tree grows
naturally near Aleppo and in several other parts of Syria. This is a tree
of middling growth in its native soil, and in England there are none of
any large size, for most of the plants that were growing here before the
year 1740 were killed by the frost that severe winter.

17 Aware of the extremely flammable nature of *Pinus halepensis* the Greek
government has proposed planting *Pinus nigra* across Evia, but given the
huge number of *Pinus halepensis* seeds left in the soil and the fires that can
be caused by *Pinus nigra*, which is itself thin barked and resinous, with a
combustible leaf litter, this seems like a doomed mission. https://www
.ekathimerini.com/multimedia/images/1194759/firs-black-pines-to-be
-planted-to-revive-evia-forests/

18 "There was a kind of dried-up reed that was very good for starting fires
with, but these grew only on the hill-top to the left of the position, and
you had to go under fire to get them. If the Fascist machine-gunners
saw you they gave you a drum of ammunition all to yourself. Generally their aim was high and the bullets sang overhead like birds, but
sometimes they crackled and chipped the limestone uncomfortably
close, whereupon you flung yourself on your face. You went on gathering reeds, however; nothing mattered in comparison with firewood."
George Orwell, *Homage to Catalonia* (1938), p. 193.

19 Stephens et al. (2007).

20 There are many other examples of fire suppression negatively affecting
trees and ecosystems: either by encouraging massive conflagrations or

by removing an essential element from a tree species' regeneration. The great anthropologist Kalyanakrishnana Sivaramakrishnan has studied the example of fire suppression under the British Raj in India, in particular the way in which the British tried to prevent all forest fires to protect sal trees and other valuable timber. Centuries of people who lived in the sal forests had burned small areas to encourage fresh growth for pasture of cattle and to control scrub, but the sal tree, one of the valuable sources of timber remaining after the comprehensive destruction of the deodar forests, was damaged and "stunted" as a Raj forestry report put it by this practice. The forestry commission therefore issued laws that instituted a blanket ban on burning—laws that were unpopular and very difficult to enforce. Where the British succeeded, sal decreased and was crowded out by other trees; one forestry commissioner ruefully discussed the ban, "which in the moister tracts weighs heavily against the very species it was designed to assist." This is because sal relies on being burned down to the ground by fire and springing up quickly from burls to outcompete other trees; it was a glorious, oft-repeating example of the ignorance and arrogance of humans being entirely foiled by the ravishing complexity of trees. See Sivaramakrishnan (1996).

21 In Spain, 85 percent of plantations are conifers (mainly pines) and 13 percent are eucalyptus; *Pinus halepensis* can and even should burn once every 20 to 30 years and of the 4,600 square miles of pines planted in Spain between 1940 and 1981, about 40 percent burned in an interval of only eight years. Between 1975 and 1990 the treed area burned in Spain exceeded the area planted, but even this led to little change; of all the forests burned in Greece, Spain, France, and Italy about one-third consists of *Pinus halepensis*, even though it constitutes only 17 percent of the forests in Greece, 7 percent of the Spanish, 4 percent of the French, and 3 percent of the Italian forests. This is partly of course because it is drought resistant, but mainly that it is fighting to come out on top. Where pines grow up to a certain level, a pine-and-fire cycle begins in which 15- to 30-year-old pines burn repeatedly. Mount Hymettus in Athens, once deforested, has now entered this cycle, and pyrogenic fire species grow up most acutely in unsettled ecosystems, places that—once cleared for farming—are now back in a time of pioneer planting. Grove and Rackham (2003).

22 Doughty (1996).

23 A good summary can be found in Borunda (2018).

24 See the conference notes of Eucalyptus 2018: *Managing Eucalyptus Plantations Under Global Changes*, available online at https://agritrop .cirad.fr/589039/1/ID589039.pdf

25 The Australian government's site gives excellent up-to-date information: https://www.agriculture.gov.au/abares/forestsaustralia/profiles /eucalypt-2016

26 Adam, *Australian Rainforests* (1992), Chapter 9; Pyne (1991), Chapter 2.

27 There is some evidence that eucalyptus may have been managed by pre-European peoples to form savannah. The botanical draftsman on board Captain Cook's expedition in 1788 wrote: "The country looked very pleasant and fertile, and the trees, quite free from underwood, appeared like plantations in a gentleman's park," which suggests an active form of management. Or does it? Studies have shown that pyrogenic plants can maintain a savannah-like environment even in a humid place; as with pine, eucalyptus takes a hit in terms of lost effort in burnt wood, but once again, because the fire will race over the undergrowth, it is very unlikely that the heat of the fire will penetrate the earth.

28 "Fire blog 6: The eucalypts will be back," Know Our Plants. https:// know.ourplants.org/fire/fire-blog-6-the-eucalypts-will-be-back/

29 "The Anglo-Saxon Chronicle: Eleventh Century," The Avalon Project, Yale Law School. https://avalon.law.yale.edu/medieval/ang11.asp

30 Adams (2024).

Chapter 4: Trees shaping air

1 Fayol et al. (1887), p. 156.

2 Oil, of course, is made from sea creatures and smaller plants. Magyar (2023).

3 *A Selection of the Geological Memoirs Contained in the Annales des Mines*, trans. H. T. de la Beche (1824), p. 206.

4 This view, held by most until 2010, has now been questioned—rather than living for 15 to 20 years only, some now suggest that there is evidence that the lycopsids lived for centuries, and that the higher growth rates suggested would be physiologically impossible. Boyce et al. (2016).

5 As seen in Trout Creek in the Rocky Mountains. Viney et al. (2019).

6 We know this because the structure of sporopollenin was deciphered recently, in 2019, and like a newly deciphered language it opened up a whole sheaf of possibilities and views into the past. (Augustin Pyramus de Candolle painstakingly isolated and described it in 1813 but it was not until 2019 that its structure was known, because even for humans it is so hard to break down.) The structure was deciphered in Cambridge, Massachusetts, but it's so hard to collect sporopollenin that Jing-Ke Weng, the scientist conducting the studies, asked his parents to send him pollen from the pitch pine, *Pinus rigida*, which is sold in bulk in China as a topping for rice cakes. The pollen was sheared mechanically with a high-energy ball mill and then degraded with a mixture of strong acid and sulfur. Even then, it was not an easy job and 50 percent remained unbroken. What broke was enough to provide a clear view of what was there, three relatively common ring structures, including the plant metabolites coumarate and naringin, the aromatic compound that gives grapefruit its smell. The ring structures—which can essentially damp down free-radical reactions—show how sporopolleinin protects cells from UV and suggested potential avenues for using sporopolleinin capsules for biomimetic structures like vaccine delivery capsules and even—some have suggested—space capsules.

7 Inuma and Lee (2024).

8 "A (Pollen-free) Sigh of Relief for Japan: The Genetics of Male Sterility in Cedar Trees," Forestry and Forest Products Research Institute. https://www.ffpri.affrc.go.jp/ffpri/en/research/results/2021/20210216press.html

9 Hu et al. (2023).

10 Ren et al. (2023).

11 Cintolesi et al. (2023).

12 Hooke et al. (1665), p. 116.

13 Bianchi (2021).

14 Margulis (1967). As Margulis herself writes in *Slanted Truths* (1997), a book of collected essays, it was "an early short for SET."

15 Ciais et al. (1997).

16 Brown (2023); Margulis and Sagan (eds.) *Slanted Truths* (1997).

17 Carbon dioxide competes with oxygen at the active site of Rubisco, so an increase in carbon dioxide concentration makes carbon fixation more

likely to occur. If oxygen reacts rather than carbon dioxide, the Rubisco active site is occupied, and the resultant molecule must be painstakingly unpicked at an energetic cost to the tree. Until a certain level therefore, increasing carbon dioxide levels increases the tree's carbon fixation efficiency. However, photorespiration is required for certain pathways, and the long-term effects on trees of the very fast spike of carbon dioxide generated recently by humans has not been characterized. See Wang (2020).

18 The BIFoR FACE experiment is one of the world's largest climate change experiments. It pumps out CO_2 into a Staffordshire woodland and measures the effect, which—broadly speaking—has been the increased growth of approximately 200-year-old oaks nearby. This is important because it it is yet another stinging indictment of the stupidity of replacing old forest with fast-growing tree plantations; a practice which, tragically and unbelievably, occasionally still happens in the name of carbon sequestration. See Norby (2024).

Chapter 5: Trees shaping fungi

1 For a quick sweep through the early world of plant-fungal interactions, see LePage et al. (1997).

2 The first interactions seem to have been mediated through complex drawings-up of the battle lines (as the origins of multicellular life). Simon et al. (1993).

3 A lovely description of this can be found in Richard Fortey's book *Close Encounters of the Fungal Kind* (2024).

4 For a detailed case study, see Carey et al. (2020).

5 Fahey et al. (2012). "Arbuscular mycorrhizal colonization of giant sequoia (*Sequoiadendron giganteum*) in response to restoration practices." *Mycologia* 104(5), 988–97.

6 Most AM roots have a suberized exodermis, which forms a permeability barrier around plant-fungus interfaces in the cortex and also protects roots from unwanted fungi. ECM short roots also have highly specialized anatomical features. Rich et al. (2017).

7 As ever, it is studying the unnatural absence of a chemical that tells us most about its use. See, for example, Su et al. (2023).

8 Shindo et al. (2018).

9 Akiyama et al. (2005).

10 For more about this see-saw of earth's elements, see Yoneyama et al. (2012).

11 This was told to me by a Syrian friend. As truffles are ECM, not AM, fungi this is confusing, but evidence has shown that some ECM fungi may respond to strigolactones too. Many interesting elements of truffle hunting are told in Alrhmoun et al. (2025).

12 Carvalhais et al. (2019).

13 For a better look at what connection really means in this case, see Giovannetti et al. (2006).

14 Karst quoted in Jones et al. (2023).

15 There are many papers giving both sides of the argument. See, for example, Karst et al. (2023).

16 Giovannetti et al. (2006).

17 Weinhold et al. (2021); Barto et al. (2012).

18 Steidinger et al. (2019).

19 Strullu-Derrien et al. (2018).

20 Nichols and Johnson, *Plants and the KT Boundary* (2008).

21 Genetic evidence from the Russulaceae suggests that ECM interactions may be more likely to develop in cases where fungi change their mode of nutrition (e.g. to different lengths of polysaccharide chain) than via changes in signaling pathways. As this nutritional change is particularly likely to occur in facultative saprotrophs (those fungi that can switch to metabolize dead wood but don't need it to complete their life-cycles) there's a strong case suggesting that the ECM habit develops where fungi feeding on dead wood move in to colonize the live tree. Smith et al. (2017); Brundrett and Tedersoo (2018).

22 Adams (2015).

23 The Crowther Lab has done much work in this area. See, for example, Anthony's "Why fungi secretly drive tree growth across Europe."

24 For more on acute oak decline see in particular the excellent website of Forest Research, which is updated regularly with new information: forestresearch.gov.uk

25 Since ash dieback became a problem, papers about potential amelioration strategies have boomed. Most have concluded that the answer can only be found on a large scale. See George et al. (2022).

Chapter 6: Trees shaping plants

Epigraph, Pablo Neruda, "The Chilean Forest," from *Memoirs* (1978).

1 Schlanger (2024).
2 This is a long and beautiful story, with the sweep and swell of ice back and forward, well told by Mitchell (2006).
3 Ludlow, *The Memoirs of Edmund Ludlow, Lieutenant-General of the Horse in the Army of the Commonwealth of England, 1625–1672*, vol. 1 (1894), p. 292.
4 Molloy et al. (2014).
5 Christenhusz et al., *Plants of the World: An Illustrated Encyclopaedia of Vascular Plants* (2017).
6 In the case of ferns like the Killarney fern this involves staying for years in tiny populations as a gametophyte—with only half the DNA needed to replicate. The fern, *Trichomonas speciosum*, lives in oak woodlands in Killarney, which have an undergrowth layer of *Quercus ilex*, the evergreen oak, and has an intrinsically low metabolism, surviving at low temperatures. Its sporophyte form has more non-photosynthetic cells, and therefore requires more light. These non-photosynthetic cells have a similar cost to the parenchyma, but clearly add nothing to aid the fern in seeking more light. Its survival appears to depend on infrequent periods of high photosynthesis when light finds its way through the trees. It also grows in Devon—in Wistman's Wood and in the valley of the rocks, in Mother Meldrum's Cave.
7 Hill et al. (2021); also Fan et al. (2023).
8 For years people thought that this was what Darwin meant when he wrote about "the abominable mystery" but more recently Richard Buggs at Kew Gardens has revealed that this is not the case. Darwin thought that monocots had diversified for years before the great flowering of the angiosperms. He was thinking of the dicots only. Darwin may have been misled, but the question remains. Why did the angiosperms suddenly explode in diversity and thus push the conifers out? Zuntini et al. (2024).
9 Carvalho et al. (2021).
10 Dimitrov et al. (2021).
11 See Crane et al. (1995).

12 Quoted from *In Search of Flowers of the Amazon Forest* by Margaret Mee (1988), pp. 292–6. The pollinator is believed never to have been caught in the act of pollinating the cactus in the wild, but because of the long stalk that supports the flower it is thought that only two known moths have a proboscis long enough to reach the nectar: *Amphimoaea walkerii* or the Darwin hawkmoth (which has the longest proboscis in the world) and *Cocytus cruentes*. It seems to me, however, almost incredible that no one in the Baré tribe has ever seen this happening. The seeds have an air chamber inside that allows them to float down the Rio Negro during the high water. For a nice overview see "Secrets of the Moonflower," Cambridge University Botanic Garden. https://www.botanic.cam.ac.uk/secrets-of-the-moonflower

13 For the details of the carbon cycle in the Chilean temperate rainforest, see Urrutia-Jalabert et al. (2015), and for Scots pine in Ireland, see Roche et al. (2009).

Chapter 7: Trees shaping animals

1 For a nuanced discussion of the various habits of sloths and the evolutionary reasons behind this, see Pauli et al. (2014).

2 For an excellent explanation of bee eyes and how they work, see Riddle (2016).

3 The same flower signal was attractive to humans, and it was because of the spikes of white candles that the horse chestnut finally spread out of its gorge refuge. From Zagori the horse chestnut found its way to Istanbul. Busbeq told Matthioli, who took it to the Czech court, and from there it spread across Europe. It became the symbol of Kyiv, and if you take a train journey across any part of Europe in spring you will see it flashing past, a white splash across the landscape. People had assumed that it came from Istanbul or further east until DNA studies revealed the truth of its survival in those humid gorges. See Lack (2001).

4 Much of the work on this was done during the 1980s. For diagrams and an excellent account, see Mustoe (2007).

5 This was recorded in August 1602 by Captain Wilhem Van Westzanen; see Staub (1995).

6 For an excellent overview, see Dantas (2022).

7 For an imaginative and evocative account of a possible Gomphothere feeding pattern, see Janzen and Martin (1982).

8 Letter to J. D. Hooker, 7 April 1874, in *A Calendar of the Correspondence of Charles Darwin, 1821–1882* (1985).

9 Joshua tree subspecies have been extensively studied. See Royer (2016).

10 There are many brilliant papers exploring this: see in particular Esque et al. (2023).

11 Previous studies of floral abscission in Yucca *filamentosa* suggest that wounding of ovules during oviposition may prompt floral abortion and that moths with longer ovipositors may be more likely to wound ovules triggering this abscission response. The longer ovipositor in *Tegeticula synthetica* vs. *Tegeticula antithetica* matches the difference in the length of the style of their respective hosts. Thus, floral abscission may occur when *T. synthetica* oviposits on *Y. brevifolia jaegeriana* but not when *T. antithetica* oviposits on *Y. brevifolia brevifolia*, which may explain the pattern of asymmetric hybridization in Joshua trees. *Yucca brevifolia jaegeriana* requires a longer ovipositor, so is more likely to tolerate the opposite species as a pollinator. Marr and Pellmyr (2003).

12 Takahashi et al. (2014).

13 This subtle scent-switch happens in many tree species, to the extent that the insect "choice" in the matter can be called into question. Wang et al. (2019).

14 These relationships can get as complicated in trees as they do in other ecological webs. See, for example, Pereira et al. (2022).

15 For a comprehensive and convincing account, see Mohanta et al. (2012).

16 As quoted in Lynn Margulis' collection of essays: Margulis and Sagan (eds.), *Slanted Truths* (1997).

17 Motamayor et al. (2002).

18 Karremans, *Demystifying Orchid Pollination* (2023), p. 160.

Chapter 8: Trees shaping people

Epigraph: Rebecca Solnit, *Orwell's Roses* (2021), p. 12.

1 This entire chapter owes much to the ideas superbly told in Enos' *The Wood Age* (2021). For further detail see Bonnefille et al. (2004).

2 DeSilva et al. (2010).

3 For a nice overview of the controversy, see Sample (2016).

4 Thorpe et al. (2006).

5 Fifty years after Lucy's discovery, *Science* magazine dedicated an edition to her and the controversies surrounding early human development. See in particular Gibbons (2024).

6 DeSilva et al. (2010).

7 Drummond-Clarke (2023).

8 Hernandez-Aguilar et al. (2013).

9 Pollan (2001) quoted in Scott (2017). For more on the fruits of the Amazon, see Prance (2015).

10 For more on cacao evolution and expansion, see Zarrillo et al. (2018) and Lanaud et al. (2024).

11 For more on vanillin and its role, see Cameron (2023).

12 While rats can generate vitamin C, primates, bony fishes, and several other groups of animals have lost the ability to synthesize it and require it in their diets. Intriguingly, this deficiency is always caused by loss of L-gulonolactone oxidase (L-GulLO), the final enzyme in the biosynthetic pathway. Duque et al. (2022).

13 For a brilliant article about this thrilling tale, see Trickey (2016).

14 Thomas Jefferson, letter to W. S. Smith, 13 November 1787, *The Papers of Thomas Jefferson, Volume 12, 7 August 1787–31 March 1788* (1955), pp. 355–7.

15 "Liberty has its roots in the heart of the people, like the tree in the heart of the earth; like the tree it raises and spreads its branches in the sky; like the tree, it grows unceasingly and covers generations with its shade. The first tree of freedom was planted, eighteen hundred years ago, by God himself on Golgotha. The first tree of liberty is this cross on which Jesus Christ offered himself as a sacrifice for the liberty, equality and fraternity of the human race." Victor Hugo, "Speech at the planting of a Tree of Liberty on the Place des Vosges," 2 March 1848. Quoted in Beecher, *Writers and Revolution: Intellectuals and the French Revolution of 1848* (2021).

16 Bensaïd and Nichols, *The Dispossessed* (2021).

17 Coughlan (2013).

18 *Vincent Van Gogh: A Life in Letters* (2020), p. 386.

Bibliography

Achatz, M., Morris, E. K., Mueller, F., Hilker and M., Rillig, M. C. (2014). "Soil hypha-mediated movement of allelochemicals: arbuscular mycorrhizae extend the bioactive zone of juglone." *Functional Ecology*, vol. 28, issue 4, pp. 1020–9.

Adam, P. (1992). *Australian Rainforests* (Oxford University Press, Oxford)

Adams, A. M. (2024). "Old-Growth Forests Know How to Protect Themselves from Fire." *Scientific American*, 19 March 2024. https://www.scientificamerican.com/article/can-forests-protect-themselves/

Adams, M. A. and Pfautsch, S. (2018). "Grand Challenges: Forests and Global Change." *Frontiers in Forests and Global Change*, vol. 1.

Agee, J. K. (1998). "Fire and pine ecosystems" in ed. Richardson, D. M., *Ecology and Biogeography of Pinus*, pp. 193–218 (Cambridge University Press, Cambridge)

Ahlgren, R. (1994). "Soil acidification and nitrogen saturation from weathering of ammonium-bearing rock." *Nature*, vol. 368, pp. 838–41.

Alrhmoun, M. et al. (2025). "Ethnoecology of desert truffles hunting: A cross-cultural comparative study on practices and perceptions in the Mediterranean and the Near East." *Journal of Arid Environments*, vol. 229, 105367.

Akiyama, K., Matsuzaki, Ki. and Hayashi, H. (2005). "Plant sesquiterpenes induce hyphal branching in arbuscular mycorrhizal fungi." *Nature*, vol. 435, pp. 824–7.

al-Malā'ikah, N. (2020). *Revolt Against the Sun: The Selected Poetry of Nāzik al-Malā'ikah, A Bilingual Reader*, ed. and trans. Drumsta, E. (Saqi Books, London)

Alban, D. H. (1969). "The influence of Western Hemlock and

Western Redcedar on soil properties." *Soil Science Society of America Journal*, vol. 33, issue 3, pp. 453–7.

Alves et al. (2016). "Seasonality of isoprenoid emissions from a primary rainforest in central Amazonia." *Atmospheric Chemistry and Physics*, vol. 16, issue 6.

Anthony, M. (n.d.). "Why fungi secretly drive tree growth across Europe." Crowther Lab. https://crowtherlab.com/why-fungi-secretly-drive-tree-growth-across-europe/

Antonelli, A., Nylander, J. A., Persson, C. and Sanmartín, I. (2009). "Tracing the impact of the Andean uplift on Neotropical plant evolution." *Proceedings of the National Academy of Sciences of the United States of America*, vol. 106, no. 24, pp. 9749–54.

Ash, S. R. and Creber, G. T. (2000). "The Late Triassic *Araucarioxylon Arizonicum* Trees of the Petrified Forest National Park, Arizona, USA." *Paleontology*, vol. 43, part 1, pp. 15–28.

Ash, S. R. and Savidge, R. A. (2004). "The Bark of the Late Triassic *Araucarioxylon Arizonicum* Tree from Petrified Forest National Park, Arizona." *IAWA Journal*, vol. 25, issue 3, pp. 349–68.

Bainbridge, D. A., (1985). "The Rise of Agriculture: A New Perspective Source." *Ambio*, vol. 14, no. 3, pp. 148–51.

Baladón, A. (1973). "Cultivos enarenados." Instituto Nacional de Meteorogia Madrid, A-55.

Baldwin, I. and Schutz, J. C. (1983). "Rapid Changes in Tree Leaf Chemistry Induced by Damage: Evidence for communication between plants." *Science*, vol. 221, issue 4607.

Balter, M. (2015). "Raging fires, high temps kept big dinosaurs out of North America for millions of years." *Science*, 15 June 2015. https://www.science.org/content/article/raging-fires-high-temps-kept-big-dinosaurs-out-north-america-millions-years

Barquera, R. et al. (2024). "Ancient genomes reveal insights into ritual life at Chichén Itzá." *Nature*, vol. 630, pp. 912–19.

Barrett, P. M. and Willis, K. J. (2001). "Did dinosaurs invent flowers? Dinosaur-angiosperm coevolution revisited." *Biological reviews of the Cambridge Philosophical Society*, vol. 76, issue 3, pp. 411–47.

Barto, E. K., Weidenhamer, J. D., Cipollini, D. and Rillig, M. C. (2012). "Fungal superhighways: do common mycorrhizal networks enhance below ground communication?" *Trends in Plant Science*, vol. 17, issue 11, pp. 633–7.

Beecher, J. (2021). *Writers and Revolution: Intellectuals and the French Revolution of 1848*. (Cambridge University Press, Cambridge)

Beerling, D. (2017). *The Emerald Planet: How Plants Changed Earth's History* (Oxford University Press, Oxford)

Belton, S. et al. (2024). "Molecular characterization of *Pinus sylvestris* (L.) in Ireland at the western limit of the species distribution." *BMC Ecology and Evolution*, vol. 24, 12.

Bensaid, D. and Nichols, R. (2021). *The Dispossessed: Karl Marx's Debates on Wood Theft and the Right of the Poor* (University of Minnesota Press, Minneapolis)

Berendse, F. and Scheffer, M. (2009). "The angiosperm radiation revisited, an ecological explanation for Darwin's 'abominable mystery.'" *Ecology Letters*, vol. 12, issue 9, pp. 865–72.

Bermúdez-Contreras, A. I., Monroy-Guzmán, C., Pérez-Lucas, L., Escutia-Sánchez, J. A., Del Olmo-Ruiz, M. and Truong, C. (2022). "Mycorrhizal Fungi Associated with Juniper and Oak Seedlings Along a Disturbance Gradient in Central Mexico." *Frontiers in Forests and Global Change*, vol. 5.

Bianchi, T. S. (2021). "The evolution of biogeochemistry: revisited." *Biogeochemistry*, vol. 154, pp. 141–81.

Bidartondo, M., Read, D. J., Trappe, J. M., Merckx, V., Ligrone R. and Duckett J. G. (2011). "The dawn of symbiosis between plants and fungi." *Biology Letters*, vol. 7, pp. 574–7.

Blonder, B., Royer, D. L., Johnson. K. R., Miller, I. and Enquist, B. J. (2014). "Plant Ecological Strategies Shift Across the Cretaceous–Paleogene Boundary." *PLoS Biology*, vol. 12, issue 9.

Boatwright, A., Hughes, S. and Barry, J. (2015). "The height limit of a siphon." *Scientific Reports*, vol. 5, 16790.

Bond, W. J. and Midgley, J. J. (2012). "Fire and the Angiosperm Revolutions." *International Journal of Plant Sciences*, vol. 173, no. 6, pp. 569–83.

Bond, W. J., Woodward, F. I. and Midgley, G. F. (2005). "The global distribution of ecosystems in a world without fire." *New Phytologist*, vol. 165, issue 2, pp. 525–38.

Bonfante, P. and Genre, A. (2010). "Mechanisms underlying beneficial plant–fungus interactions in mycorrhizal symbiosis." *Nature Communications*, vol. 1, issue 48.

Bonnefille, R. et al. (2004). "High-resolution vegetation and climate change associated with Pliocene *Australopithecus afarensis*." *Proceedings of the National Academy of Sciences of the USA*, vol. 101, no. 33, pp. 12125–9.

Borunda, A. (2018). "Koalas Eat Toxic Leaves to Survive: Now Scientists Know How." *National Geographic*, 2 July 2018. https://www.nationalgeographic.com/animals/article/scientists-sequenced-the-koala-genome-to-save-them

Bowles, A. M. C., Paps, J. and Bechtold, U. (2022). "Water-related innovations in land plants evolved by different patterns of gene cooption and novelty." *The New Phytologist*, vol. 235, issue 2, pp. 732–42.

Boyce, C. K., Brodribb, T. J., Feild, T. S. and Zwieniecki, M. A. (2009). "Angiosperm leaf vein evolution was physiologically and environmentally transformative." *Proceedings of the Royal Society B: Biological Sciences*, vol. 276, pp. 1771–6.

Boyce, C. K. and DiMichele, W. A. (2016). "Arborescent lycopsid productivity and lifespan: Constraining the possibilities." *Review of Paleobotany and Palynology*, vol. 227, 97–110.

Boyce, C. K., Fan, Y. and Zwieniecki, M. A. (2017). "Did trees grow up to the light, up to the wind, or down to the water? How modern high productivity colors perception of early plant evolution." *New Phytologist*, vol. 215, issue 2, pp. 552–7.

Boyce, C. K. and Lee, J. E. (2017). "Plant Evolution and Climate Over Geological Timescales." *Annual Review of Earth and Planetary Sciences*, vol. 45, pp. 61–87.

Boyce, C. K., Lee, J. E., Feild, T. S., Brodribb, T. and Zwieniecki, M. A. (2010). "Angiosperms helped put the rain in the rainforests: The impact of plant physiological evolution on tropical biodiversity." *Annals of the Missouri Botanical Garden*, vol. 97, pp. 527–40.

Bramwell, Z. and Bramwell, D. (1974). *Wild Flowers of the Canary Islands* (Stanley Thornes Ltd., London and Burford)

Briand, C. H. (2000). "Cypress knees: An enduring enigma." *Arnoldia*, vol. 60, issue 4, pp. 19–25.

Brillouet, J. M., Romieu, C., Schoefs, B., Solymosi, K., Cheynier, V., Fulcrand, H., Verdeil, J. L. and Conéjéro, G. (2013). "The tannosome is an organelle forming condensed tannins in the chlorophyllous organs of Tracheophyta." *Annals of Botany*, vol. 112, issue 6, pp. 1003–14.

Bringmann, M. et al., (2012). "Cracking the elusive alignment hypothesis: the microtubule–cellulose synthase nexus unraveled." *Trends in Plant Science*, vol. 17, no. 11.

Brodribb T. J., Field T. S. and Jordan G. J. (2007). "Leaf Maximum Photosynthetic Rate and Venation Are Linked by Hydraulics." *Plant Physiology*, vol. 144, pp. 1890–8.

Brodribb T. J. and Hill, R. S. (1997). "Imbricacy and Stomatal Wax Plugs Reduce Maximum Leaf Conductance in Southern Hemisphere Conifers." *Australian Journal of Botany*, vol. 45, issue 4, pp. 657–68.

Brodribb, T. J. and Hill, R. S. (1998). "The photosynthetic drought physiology of a diverse group of southern hemisphere conifer species is correlated with minimal seasonal rainfall." *Functional Ecology*, vol. 12, issue 3, pp. 465–71.

Brodribb, T. J., Pittermann, J. and Coomes, D. A. (2012). "Elegance versus Speed: Examining the Competition Between Conifer and Angiosperm Trees." *International Journal of Plant Sciences*, vol. 173, No. 6, pp. 673–94.

Brown, D. W. (2023). "A Security Camera for the Planet." *New Yorker*, 28 April 2023. https://www.newyorker.com/news/annals-of-climate-action/a-security-camera-for-the-planet

Brundrett, M. C. (2002). "Coevolution of roots and mycorrhizas of land plants." *New Phytologist*, vol. 154, pp. 275–304.

Brundrett, M. C. and Tedersoo, L. (2018). "Evolutionary history of mycorrhizal symbioses and global host plant diversity." *New Phytologist*, vol. 220, pp. 1108–15.

Brunoud, G., Wells, D. M., Oliva, M., Larrieu, A., Mirabet, V.,

Burrow, A. H., Beeckman, T., Kepinski, S., Traas, J. and Bennett, M. J. (2012). "A novel sensor to map auxin response and distribution at high spatio-temporal resolution." *Nature*, vol. 482, pp. 103–6.

Buatois, L. A., Davies, N. S., Gibling, M. R., Krapovickas, V., Labandeira, C. C., MacNaughton, R. B., Mángano, M. G., Minter, N. J. and Shillito, A. P. (2022). "The Invasion of the Land in Deep Time: Integrating Paleozoic Records of Paleobiology, Ichnology, Sedimentology, and Geomorphology." *Integrative and Comparative Biology*, vol. 62, issue 2, pp. 297–331.

Buras, A., Thevs, N., Zerbe, S. and Wilmking, M. (2013). "Productivity and carbon sequestration of *Populus euphratica* at the Amu River, Turkmenistan." *Forestry: An International Journal of Forest Research*, vol. 86, issue 4, pp. 429–39.

Burton, J. E. et al. (2020). "Leaf traits predict global patterns in the structure and flammability of forest litter beds." *Journal of Ecology*, vol. 109, issue 3, pp. 1344–55.

Byers, B. A., DeSoto, L., Chaney, D. et al. (2020), "Fire-scarred fossil tree from the Late Triassic shows a pre-fire drought signal." *Scientific Reports*, vol. 10, 20104.

Caetano Andrade, V. L., Flores, B. M., Levis, C., Clement, C. R., Roberts, P. and Schöngart, J. (2019). "Growth rings of Brazil nut trees (*Bertholletia excelsa*) as a living record of historical human disturbance in Central Amazonia." *PLoS ONE*, vol. 14, issue 4.

Callcott, M. (1993). *The Captain's Wife: The South American Journals of Maria Graham, 1821–1823*, ed. Mavor. E. (Weidenfeld & Nicolson, London)

Camargo, M. A. B. and Marenco, R. A. (2010). "Density, size and distribution of stomata in 35 rainforest tree species in Central Amazonia." *Acta Amazonica*, vol. 41, issue 2, pp. 205–12.

Cameron, K. M. (2023). "Plant ecology: Vanilla lures both insects and mammals to disperse its seeds and fruits." *Current Biology*, vol. 33, issue 2.

Campbell, D. G. (2005). *A Land of Ghosts: The Braided Lives of People and the Forest in Far Western Amazonia* (Houghton Mifflin Harcourt, New York)

Cao, L. et al. (2023). "Floral scent of the Mediterranean fig tree:

significant inter-varietal difference but strong conservation of the signal responsible for pollinator attraction." *Scientific Reports*, vol. 13, 5642.

Cardoso, D., Carvalho-Sobrinho, J., Zartman, C., Komura, D. and Queiroz, L. (2015). "Unexplored Amazonian diversity: Rare and phylogenetically enigmatic tree species are newly collected." *Neodiversity*, vol. 8, pp. 55–73.

Carey, C. J., Glassman, S. I., Bruns, T. D., Aronson, E. L. and Hart, S. C. (2020). "Soil microbial communities associated with giant sequoia: How does the world's largest tree affect some of the world's smallest organisms?" *Ecology and Evolution*, vol. 10, issue 13, pp. 6593–609.

Carvalhais, L. C., Rincon-Florez, V. A., Brewer, P. B., Beveridge, C. A., Dennis, P. G. and Schenk, P. M. (2019). "The ability of plants to produce strigolactones affects rhizosphere community composition of fungi but not bacteria." *Rhizosphere*, vol. 9, pp. 18–26.

Carvalho, M. R. et al., (2021). "Extinction at the end-Cretaceous and the origin of modern Neotropical rainforests." *Science*, vol. 372, no. 6537, pp. 63–8.

Chen, C. et al (2009). "Private channel: a single unusual compound assures specific pollinator attraction in *Ficus semicordata*." *Functional Ecology*, vol. 23, pp. 941–50.

Chen, H. and Markham, J. (2021). "Ancient CO_2 levels favor nitrogen fixing plants over a broader range of soil N compared to present." *Scientific Reports*, vol. 11, issue 1, 3038.

Choat, B., et al. (2018). "Triggers of tree mortality under drought." *Nature*, vol. 558, pp. 531–9.

Christenhusz, M. J. M., Fay, M. F. and Chase, M. W. (2017). *Plants of the World: An Illustrated Encyclopaedia of Vascular Plants* (Kew Publishing & University of Chicago Press, Richmond, Surrey and Chicago)

Ciais, P., et al. (1997). "A three-dimensional synthesis study of $\delta^{18}O$ in atmospheric CO_2. 1. Surface fluxes." *Journal of Geophysical Research: Atmospheres*, vol. 102, pp. 5857–72.

Cintolesi, C. et al. (2023). "Characterization of flow dynamics within and around an isolated forest, through measurements and numerical simulations." *Agricultural and Forest Meteorology*, vol. 339, 109557.

Conti, E., Franks, N. P. and Brick, P. (1996). "Crystal structure of firefly luciferase throws light on a superfamily of adenylate-forming enzymes." *Structure*, vol. 4, issue 3, pp. 287–98.

Coughlan, M. R. (2013). "Errakina: Pastoral Fire Use and Landscape Memory in the Basque Region of the French Western Pyrenees." *Journal of Ethnobiology*, vol. 33, issue 1, pp. 86–104.

Crane, P., Friis, E. and Pedersen, K. (1995). "The origin and early diversification of angiosperms." *Nature*, vol. 374, pp. 27–33.

Crisp, M. D. and Cook, L. G. (2013). "How Was the Australian Flora Assembled Over the Last 65 Million Years? A Molecular Phylogenetic Perspective." *Annual Review of Ecology, Evolution, and Systematics*, vol. 44, pp. 303–24.

Cunha, H. F. V., Andersen, K. M., Lugli, L. F. et al. (2022). "Direct evidence for phosphorus limitation on Amazon forest productivity." *Nature*, vol. 608, pp. 558–62.

Da Costa, D. P., Peralta, D. F., Buck, W. R., Larrain, J. and Von Konrat, M. (2017). "Serra Do Curicuriari, Amazonas State, Brazil: The First Bryofloristic Analysis for a Brazilian Mountain in the Amazonian Forest." *Phytotaxa*, vol. 303, issue 3, pp. 201–17.

Dahl, T. W. and Arens, S. K. M. (2020). "The impacts of land plant evolution on Earth's climate and oxygenation state: An interdisciplinary review." *Chemical Geology*, vol. 547, 119665.

Dantas, V. L. and Pausas, J. G. (2022). "The legacy of the extinct Neotropical megafauna on plants and biomes." *Nature Communications*, vol. 13, 129.

Darwin, C. (1985). "Letter to J. D. Hooker, 7 April 1874," in *A Calendar of the Correspondence of Charles Darwin, 1821–1882*, eds. Burkhardt, F. and Smith, S. (Cambridge University Press, Cambridge)

Darwin, C. and Darwin, F. E. (1880). "Sensitiveness of Plants to Light: It's Transmitted Effect." *The Power of Movement in Plants*, pp. 574–92. (John Murray, London)

Davies, N. S., Berry, C. M., Marshall, J. E. A. et al. (2021). "The Devonian landscape factory: plant–sediment interactions in the

Old Red Sandstone of Svalbard and the rise of vegetation as a biogeomorphic agent." *Journal of the Geological Society*, vol. 178, issue 5, jgs2020-225.

Davies, N. S., McMahon, W. J. and Berry, C. M. (2024). "Earth's earliest forest: fossilized trees and vegetation-induced sedimentary structures from the Middle Devonian (Eifelian) Hangman Sandstone Formation, Somerset and Devon, SW England." *Journal of the Geological Society*, vol. 181, issue 4.

Davis, W. (1997). *One River: Explorations and Discoveries in the Amazon Rain Forest* (Simon & Schuster, London)

de Andrade, M. (2023), *Macunaíma*, trans. Dodson, K. (Fitzcarraldo, London)

De La Torre, A. R., Piot, A., Liu, B., Wilhite, B., Weiss, M. and Porth, I. (2019). "Functional and morphological evolution in gymnosperms: A portrait of implicated gene families." *Evolutionary Applications*, vol. 13, pp. 210–27.

de Witte, L. C. and Stocklin, J. (2010). "Longevity of clonal plants: why it matters and how to measure it." *Annals of Botany*, vol. 106, pp. 859–70.

Decombeix, A., Meyer-Berthaud, B. and Galtier, J. (2011). "Transitional changes in arborescent ligniophytes at the Devonian-Carboniferous boundary." *Journal of the Geological Society*, vol. 168, pp. 547–57.

Delclòs, X. et al. (2023). "Amber and the Cretaceous Resinous Interval." *Earth-Science Reviews*, vol. 243, 104486.

DeSilva, J. M. and Throckmorton, Z. J. (2010). "Lucy's flat feet: the relationship between the ankle and rearfoot arching in early hominins." *PLoS ONE*, vol. 5, issue 12.

Detlefsen, K. (2018). "Émilie du Châtelet," *Stanford Encyclopedia of Philosophy*, Winter 2018 edition, ed. Zalta, E. N. https://plato.stanford.edu/archives/win2018/entries/emilie-du-chatelet/

Dijkstra, F. A., Jenkins, M., de Rémy de Courcelles, V., Keitel,C., Barbour, M. M., Kayler, Z. E. and Adams, M. A. (2017). "Enhanced decomposition and nitrogen mineralization sustain rapid growth of *Eucalyptus regnans* after wildfire." *Journal of Ecology*, vol. 105, no. 1, pp. 229–36.

Dimitrov, D. et al. (2023). "Diversification of flowering plants in space and time." *Nature Communications*, vol. 14, 7609.

Doughty, R. (1996). "Not a koala in sight: promotion and spread of eucalyptus." *Ecumene*, vol. 3, issue 2, pp. 200–14.

Drummond-Clarke, R.C. (2023). "Bringing trees back into the human evolutionary story: recent evidence from extant great apes." *Communicative & Integrative Biology*, vol. 16, issue 1, 2193001.

du Châtelet, E. (1744). *Dissertation sur la nature et la propagation du feu.* Project Gutenberg. https://www.gutenberg.org/cache/epub /73279/pg73279-images.html

— (2009). *Selected Philosophical and Scientific Writings*, ed. Zins-ser, J. P., trans. Bour, I. (University of Chicago Press, Chicago)

Ducke, A. and Black, G. A. (1953). "Phytogeographical Notes on the Brazilian Amazon." *Anais Da Academia Brasileira De Ciencias*, vol. 25, issue 1.

Duque, P., Vieira, C. P., Bastos, B. et al. (2022). "The evolution of vitamin C biosynthesis and transport in animals." *BMC Ecology and Evolution*, vol. 22, 84.

Ennos, R. (2021). *The Wood Age: How One Material Shaped the Whole of Human History* (William Collins, London)

Esque, T. C. et al. (2023). "Unprecedented distribution data for Joshua trees (*Yucca brevifolia* and *Y. jaegeriana*) reveal contemporary climate associations of a Mojave Desert icon." *Frontiers in Ecology and Evolution*, vol. 11, 1266892.

Fahey, C., York, R. A. and Pawlowska, T. E. (2012). "Arbuscular mycorrhizal colonization of giant sequoia (*Sequoiadendron giganteum*) in response to restoration practices." *Mycologia*, vol. 104, issue 5, pp. 988–97.

Fan, J. et al. (2018). "Substantial convection and precipitation enhancements by ultrafine aerosol particles." *Science*, vol. 359, issue 6374, pp. 411–18.

Fan, X., Hao, X., Zhang, S. et al. (2023). "*Populus euphratica* counteracts drought stress through the dew coupling and root hydraulic redistribution processes." *Annals of Botany*, vol. 131, issue 3, pp. 451–61.

Farjon, A. (2017). *A Handbook of the World's Conifers*, 2 vols. (Brill, Leiden)

— (2018). "The Kew Review: Conifers of the World." *Kew Bulletin*, vol. 73, 8.

— (2001). *World Checklist and Bibliography of Conifers*, 2nd ed. (Kew Publishing, London)

Fayol, H., Renault, B., Zeiller, R., Brongniart, Ch. and Sauvage, H. É. (1887). "Études sur le terrain houiller de Commentry." Imprimerie Théolier. https://www.biodiversitylibrary.org/item/286171

Field, T. S., Arens, N. C., Doyle, J. A., Dawson T. E. and Donoghue M. J. (2004). "Dark and disturbed: a new image of early angiosperm ecology." *Paleobiology*, vol. 30, issue 1, pp. 82–107.

Figueroa-Rangel, B. L., Willis, K. J. and Olvera-Vargas, M. (2008). "4200 years of pine-dominated upland forest dynamics in west-central Mexico: Human or natural legacy?" *Ecology*, vol. 89, issue 7, pp. 1893–907.

Fletcher, M-S., Hall, T. and Alexandra, A. N. (2021). "The loss of an indigenous constructed landscape following British invasion of Australia: An insight into the deep human imprint on the Australian landscape." *Ambio*, vol. 50, pp. 138–149.

Flores, B. M., Montoya, E., Sakschewski, B. et al. (2024). "Critical transitions in the Amazon forest system." *Nature*, vol. 626, pp. 555–64.

Forde, B. G. and Lea, P. J. (2007). "Glutamate in plants: metabolism, regulation, and signaling." *Journal of Experimental Botany*, vol. 58, no. 9, pp. 2339–58.

Fortey, R. (2024). *Close Encounters of the Fungal Kind: In Pursuit of Remarkable Mushrooms* (William Collins, London)

Frasier, C. L., Albert, V. A. and Struwe, L. (2008). "Amazonian lowland, white sand areas as ancestral regions for South American biodiversity: Biogeographic and phylogenetic patterns in *Potalia* (Angiospermae: Gentianaceae)." *Organisms, Diversity & Evolution*, vol. 8, issue 1, pp. 44–57.

Freitas, S. R. and Longo, K. M. (2005). "Emissões de queimadas em ecossistemas da América do Sul." *Estudos Avançados*, vol. 19, no. 53, pp. 167–85.

Gagnon, P. R., Passmore, H. A., Platt, W. J., Myers, J. A., Paine, C. E. and Harms, K. E. (2010). "Does pyrogenicity protect burning plants?" *Ecology*, vol. 91, issue 12, pp. 3481–514.

Garnica-Díaz, C. et al. (2023). "Global Plant Ecology of Tropical Ultramafic Ecosystems." *Botanical Review*, vol. 89, pp. 115–57.

George, J. P., Sanders, T. G. M., Timmermann, V. et al. (2022). "European-wide forest monitoring substantiate the necessity for a joint conservation strategy to rescue European ash species (*Fraxinus* spp.)." *Scientific Reports*, vol. 12, 4764.

Gibbons, A. (2024). "Lucy's World: Was Lucy the mother of us all? Fifty years after her discovery, the 3.2-million-year-old skeleton has rivals." *Science*, 4 April 2024. https://www.science.org/content/article/was-lucy-mother-us-all-fifty-years-discovery-famed-skeleton-rivals

Giovannetti, M., Avio, L., Fortuna, P., Pellegrino, E., Sbrana, C. and Strani, P. (2006). "At the root of the wood wide web: self recognition and non-self incompatibility in mycorrhizal networks." *Plant Signaling & Behavior*, vol.1, issue 1, pp. 1–5.

Glasspool, I. J. and Scott, A. C. (2010). "Phanerozoic concentrations of atmospheric oxygen reconstructed from sedimentary charcoal." *Nature Geoscience*, vol. 3, issue 9, pp. 627–30.

Gross, R. (2013). "Fellowship of the Tree Rings." *National Geographic*, 24 April 2013.

Grove, A. T. and Rackham, O. (2003). *The Nature of Mediterranean Europe: An Ecological History* (Yale University Press, New Haven)

Halbwachs, H., Easton, G. L., Bol, R., Hobbie, E. A., Garnett, M. H., Peršoh, D., Dixon, L., Ostle, N. J., Karasch, P. and Griffith, G. (2018). "Isotopic evidence of biotrophy and unusual nitrogen nutrition in soil-dwelling Hygrophoraceae." *Environmental Microbiology*, vol. 20, issue 10, pp. 3573–88.

Hayward, B. W. (1989). *Kauri Gum and the Gumdiggers* (The Bush Press, New Zealand)

He, T., Pausas, J. G., Belcher, C. M., Schwilk, D. W. and

Lamont, B. B. (2012). "Fire-adapted traits of *Pinus* arose in the fiery Cretaceous." *New Phytologist*, vol. 194, issue 3, pp. 751–9.

Heldt, H. and Piechulla, B. (2011). *Plant Biochemistry*, 4th ed. (Elsevier, Cambridge)

Hemming, J. (2004). *Red Gold: The Conquest of the Brazilian Indians* (Pan Books, London)

Hernandez-Aguilar, R. A., Moore, J. and Stanford, C. B. (2013). "Chimpanzee nesting patterns in savanna habitat: Environmental influences and preferences." *American Journal of Primatology*, vol. 75, pp. 979–94.

Hill, A. J., Dawson, T. E., Dody, A. and Rachmilevitch, S. (2021). "Dew water-uptake pathways in Negev Desert plants: a study using stable isotope tracers." *Oecologia*, vol. 196, pp. 353–61.

Hill, R. S. and Brodribb, T. J. (1999). "Southern conifers in time and space." *Australian Journal of Botany*, vol. 47, issue 5, pp. 639–96.

Hoogar, R. et al. (2019). "Impact of eucalyptus plantations on ground water and soil ecosystem in dry regions." *Journal of Pharmacognosy and Phytochemistry*, vol. 8.

Hooke, R., Allestry, J. and John, M. (1665). *Micrographia, or, Some physiological descriptions of minute bodies made by magnifying glasses: with observations and inquiries thereupon.* Printed by Jo. Martyn and Ja. Allestry, printers to the Royal Society. https://www.biodiversitylibrary.org/item/15485

Houlton, B. Z. et al. (2018). "Convergent evidence for widespread rock nitrogen sources in Earth's surface environment." *Science*, vol. 360, pp. 58–62.

Hu, C. et al. (2023). "Pollen in the Baltic Sea as viewed from space." *Remote Sensing of Environment*, vol. 284, 113337.

Huaraca Huasco, W., Riutta, T., Girardin, C. A. J. et al. (2021). "Fine root dynamics across pantropical rainforest ecosystems." *Global Change Biology*, vol. 27, pp. 3657–80.

Husain, I. (2012). *Basti*, trans. Prichett, F. (NYRB Classics, New York)

Immerwahr, D. (2024). "Mother trees and socialist forests: Is the wood-wide web a fantasy?" *Guardian*, 23 April 2024. https://

www.theguardian.com/environment/2024/apr/23/mother
-trees-and-socialist-forests-is-the-wood-wide-web-a-fantasy

Inuma, J. M. and Lee, M. Y. H. (2024). "Japan's answer to seasonal allergies: A subsidized tropical escape." *Washington Post*,
4 April 2024. https://www.washingtonpost.com/world/2024/04
/04/japan-hayfever-season/

Isnard, S., L'Huillier, L., Rigault, F. and Jaffré, T. (2016). "How
did the ultramafic soils shape the flora of the New Caledonian
hotspot?" *Plant and Soil*, vol. 403, pp. 53–76.

Janzen, D. H. and Martin, P. S. (1982). "Neotropical Anachro-
nisms: The Fruits the Gomphotheres Ate." *Science*, vol. 215.

Jefferson, T. (1955). *The Papers of Thomas Jefferson, Volume 12, 7 August 1787–31 March 1788*, ed. Boyd, J. P. et al. (Princeton University Press, Princeton)

Jones, M. Hoeksema, J. and Karst, J. (2023). "Where the 'Wood-
Wide Web' Narrative Went Wrong." *Undark*, 25 May 2023.
https://undark.org/2023/05/25/where-the-wood-wide-web
-narrative-went-wrong/

Jura-Morawiec, J. and Marcinkiewicz, J. (2020). "Wettability,
water absorption and water storage in rosette leaves of the
dragon tree (*Dracaena draco* L.)." *Planta*, vol. 252, 30.

Karremans, A. P. (2023). *Demystifying Orchid Pollination* (Kew Publishing, London)

Karst, J., Jones, M. D. and Hoeksema, J. D. (2023). "Positive Cita-
tion Bias and Overinterpreted Results Lead to Misinformation
on Common Mycorrhizal Networks in Forests." *Nature Ecology
and Evolution*, pp. 501–11.

Keeley, J. E. et al (2011). "Fire as an evolutionary pressure shaping
plant traits." *Trends in Plant Science*, vol. 16, issue 8, pp. 406–11.

Kenrick, P. and Strullu-Derrien, C. (2014). "The Origin and Early
Evolution of Roots." *Plant Physiology*, vol. 166, pp. 570–80.

Khatoon, A., Rehman, S. U., Aslam, M. M., Jamil, M. and Kom-
atsu, S. (2020). "Plant-Derived Smoke Affects Biochemical
Mechanism on Plant Growth and Seed Germination." *International Journal of Molecular Sciences*, vol. 21, 7760.

Kirch, P. V. (2012). *A Shark Going Inland Is My Chief: The*

Island Civilization of Ancient Hawai'i (University of California Press, Oakland)

Klimko, M., Nowińska, R., Wilkin, P. and Wiland-Szymańska,
J. (2018). "Comparative leaf micromorphology and anatomy of
the dragon tree group of *Dracaena* (Asparagaceae) and their taxonomic implications." *Plant Systematics and Evolution*, vol. 304,
pp. 1041–55.

Klimko, M. and Wiland-Szymańska, J. (2008). "Scanning electron
microscopic studies of leaf surface in taxa of genus *Dracaena* L.
(Dracaenaceae)." *Botanika—Steciana*, vol. 12, pp. 117–27.

Kneip, C., Lockhart, P., Voss, C. et al. (2007). "Nitrogen fixation
in eukaryotes: New models for symbiosis." *BMC Evolutionary
Biology*, vol. 7, 55.

Koller, J., Baumer, U., Kaup, Y., Schmid, M. and Weser, U. (2023).
"Analysis of a pharaonic embalming tar." *Nature*, vol. 425.

Körner, C. (1988). "Does global increase of CO_2 alter stomatal
density?" *Flora*, vol. 181, issues 3–4, pp. 253–7.

Lack, H. W. (2001). "Lilac and horse-chestnut: discovery and re-
discovery." *Bocconea*, vol. 13, pp. 613–6.

Lack, H. W. and Martius, C. F. P. (2022). *The Book of Palms*
(Taschen, Cologne)

Lamant, T. (2012). "Vegetative reproduction in gymnosperms." *Journal de l'Association des Parcs Botaniques de France*, n.
53, 5 pp. Does not contain specific citations but bibliography
cites Adams (2008), Burrows et al. (2003), Evans (1982), Farjon
(2005), Matthews (1991), and Zegers (2006).

Lanaud, C., Vignes, H., Utge, J. et al. (2024). "A revisited history
of cacao domestication in pre-Columbian times revealed by ar-
chaeogenomic approaches." *Scientific Reports*, vol. 14, 2972.

Lane, N. (2002). *Life Ascending: The Ten Great Inventions of Evolution*
(Profile Books, London)

— (2009). *Oxygen: The Molecule That Made the World* (Oxford University Press, Oxford)

— (2005). *Power, Sex, Suicide: Mitochondria and the Meaning of Life*
(Oxford University Press, Oxford)

Latour, B. and Lenton, T. M. (2019). "Extending the Domain of

Freedom, or Why Gaia Is So Hard to Understand." *Critical Enquiry*, vol. 45, no. 3, pp. 659–80.

Lenz, L. W. (2007). "Reassessment of *Yucca brevifolia* and Recognition of *Y. jaegeriana* as a Distinct Species." *Aliso: A Journal of Systematic and Floristic Botany*, vol. 24, issue 1.

LePage, B. A., Currah, R. S., Stockey, R. A. and Rothwell, G. W. (1997). "Fossil Ectomycorrhizae from the Middle Eocene." *American Journal of Botany*, vol. 84, issue 3, pp. 410–12.

Leys, B., Finsinger, W. and Carcaillet, C. (2014). "Historical range of fire frequency is not the Achilles" heel of the Corsican black pine ecosystem." *Journal of Ecology*, vol. 102, no. 2, pp. 381–95.

Li, L., Li, J., Rohwer, J. G., van der Werff, H., Wang, Z. H. and Li, H. W. (2011). "Molecular phylogenetic analysis of the *Persea* group (Lauraceae) and its biogeographic implications on the evolution of tropical and subtropical Amphi-Pacific disjunctions." *American Journal of Botany*, vol. 98, issue 9, pp. 1520–36.

Linebaugh, P. (2013). "Karl Marx, the Theft of Wood, and Working-Class Composition: A Contribution to the Current Debate." *Social Justice*, vol. 40, no. 1/2 (131–132), 40th Anniversary Issue: Legacies of Radical Criminology in the United States, pp. 137–61.

Looney, B., Miyauchi, S., Morin, E. et al. (2022). "Evolutionary transition to the ectomycorrhizal habit in the genomes of a hyperdiverse lineage of mushroom-forming fungi." *New Phytologist*, vol. 233, pp. 2294–309.

Lovelock, J. (1979). *Gaia: A New Look at Life on Earth* (Oxford University Press, Oxford)

Ludlow, E. (1894). *The Memoirs of Edmund Ludlow, Lieutenant-General of the Horse in the Army of the Commonwealth of England, 1625–1672*, vol. 1, ed. Firth, C. H. (Clarendon Press, Oxford)

Lusk, Chris. (2001). "Leaf life spans of some conifers of the temperate forests of South America." *Revista Chilena de Historia Natural*, vol. 74, 711–18.

Lyons, P. C., Mastalerz, M. and Orem, W. H. (2009). "Organic geochemistry of resins from modern *Agathis australis* and Eocene resins from New Zealand: Diagenetic and taxonomic implications." *International Journal of Coal Geology*, vol. 80, issue 1, pp. 51–62.

McElwain, J. C., Willis, K. J. and Lupia, R. (2005). "Cretaceous CO_2 decline and the radiation and diversification of angiosperms" in eds. Ehleringer, J. R., Cerling, T. E. and Dearing, M. D., *A History of Atmospheric CO_2 and Its Effect on Plants, Animals, and Ecosystems*, pp. 133–65 (Springer, New York)

McGeever, A. H. and Mitchell, F. J. G. (2016). "Re-defining the natural range of Scots Pine (*Pinus sylvestris* L.): a newly discovered microrefugium in western Ireland." *Journal of Biogeography*, vol. 43, issue 11, pp. 2199–208.

McNeill, J. R. (2004). "Woods and Warfare in World History." *Environmental History*, vol. 9, no. 3, pp. 388–410.

Maděra P. et al. (2020). "What We Know and What We Do Not Know About Dragon Trees?" *Forests*, vol. 11, issue 2, 236.

Magyar, J. (2023). "Can oil from trees be the next big thing in sustainable energy?" *Forbes*, 20 June 2023. https://www.forbes.com/sites/sap/2023/06/20/can-oil-from-trees-be-the-next-big-thing-in-sustainable-energy/

Makarieva, A. M. and Gorshkov, V. G. (2006). "Biotic pump of atmospheric moisture as driver of the hydrological cycle on land." *Hydrology and Earth System Sciences*, vol. 11, issue 2, pp. 1013–33.

Margalef, O., Sardans, J., Fernández-Martínez, M. et al. (2017). "Global patterns of phosphatase activity in natural soils." *Scientific Reports*, vol. 7, 1337.

Margulis, L. (authored as Sagan, L.) (1967). "On the origin of mitosing cells." *Journal of Theoretical Biology*. vol. 14, issue 3, pp. 225–IN6.

Margulis, L. and Sagan, D. (eds.) (1997). *Slanted Truths: Essays on Gaia, Symbiosis and Evolution* (Copernicus, New York)

Marr, D. L. and Pellmyr, O. (2003). "Effect of Pollinator-Inflicted Ovule Damage on Floral Abscission in the Yucca-Yucca Moth Mutualism: The Role of Mechanical and Chemical Factors." *Oecologia*, vol. 136, no. 2, pp. 236–43.

Mathesius, U. (2022). "Humboldt Review: Are legumes different? Origins and consequences of evolving nitrogen fixing symbioses." *Journal of Plant Physiology*, vol. 276, 153765.

Mecke, R., Mille, C. and Engels, W. (2005). "*Araucaria* beetles worldwide: evolution and host adaptations of a multigenus phytophagous guild of disjunct Gondwana-derived biogeographic occurrence." Pró Araucária Online 1: 1–18.

Mee, M. (1988). *In Search of Flowers of the Amazon Forest* (Nonesuch Editions, London)

Meesters A. G. C. A., Dolman, A. J. and Bruijnzeel, L. A. (2007). "Comment on 'Biotic pump of atmosphericmoisture as driver of the hydrological cycle on land' by A. M. Makarieva and V. G. Gorshkov." *Hydrology and Earth System Sciences*, vol. 11, pp. 1013–33.

Mguni, S. (2009). "Natural and Supernatural Convergences," *Current Anthropology*, vol. 50, no. 1, pp. 139–48.

Mitchell, F. J. G. (2006). "Where did Ireland's trees come from?" *Biology and Environment: Proceedings of the Royal Irish Academy*, vol. 106B, no. 3, pp. 251–9.

Mohanta, T. K., Occhipinti, A., Atsbaha Zebelo, S., Foti, M., Fliegmann, J. et al. (2012). "*Ginkgo biloba* Responds to Herbivory by Activating Early Signaling and Direct Defenses." *PLoS ONE*, vol. 7, issue 3.

Molloy, K. and O'Connell, M. (2014). "Post-glaciation plant colonization of Ireland: fresh insights from An Loch Mór, Inis Oírr, western Ireland." *The Irish Naturalists' Journal*, vol. 33, pp. 66–88.

Morehart, C. T., Lentz, D. L., Prufer, K. M. (2005). "Wood of the Gods: The Ritual Use of Pine (*Pinus* spp.) by the Ancient Lowland Maya." *Latin American Antiquity*, vol. 16, no. 3, pp. 255–74.

Motamayor, J., Risterucci, A., Lopez, P. et al. (2002). "Cacao domestication I: the origin of the cacao cultivated by the Mayas." *Heredity*, vol. 89, pp. 380–6.

Mustoe, G. (2007). "Coeveolution of dinosaurs and cycads." *The Cycad Newsletter*, vol. 30, no. 1.

Nelsen, M. P., DiMichele, W. A., Peters, S. E. and Boyce, C. K. (2016). "Delayed fungal evolution did not cause the Paleozoic peak in coal production." *Proceedings of the National Academy of Sciences of the USA*, vol. 113, no. 9. pp. 2442–7.

Neruda, P. (2017). *Memoirs*, trans. St. Martin, H. (Souvenir Press, London)

Newman, B. (1998). *Voice of the Living Light: Hildegard of Bingen and Her World* (University of California Press, Berkeley)

Nichols, D. J. and Johnson, K. R. (2008). *Plants and the KT Boundary* (Cambridge University Press, Boston)

Nicholls, D. G. and Ferguson, S. J. (2013). *Bioenergetics 4* (Elsevier, London)

Niklasson, M., Zin, E., Zielonka, T., Feijen, M., Korczyk, A. F., Churski, M., Samojlik, T., Jędrzejewska, B., Gutowski, J. M. and Brzeziecki, B. (2010). "A 350-year tree-ring fire record from Białowieża Primeval Forest, Poland: Implications for Central European lowland fire history." *Journal of Ecology*, vol. 98, pp. 1319–29.

Nitsiakos, V. (ed.) (2017). *Peklari: Social Economy in a Greek Village* (Lit Verlag, Münster)

Nogué, S., de Nascimento, L., Fernandez-Palacios M., Whittaker R. J. and Willis K. J. (2013). "The ancient forests of La Gomera, Canary Islands, and their sensitivity to environmental change." *Journal of Ecology*, vol. 101, issue 2, pp. 368–77.

Norby, R. J., Loader, N. J., Mayoral, C. et al. (2024). "Enhanced woody biomass production in a mature temperate forest under elevated CO_2." *Nature Climate Change*, vol. 14, pp. 983–8.

Normile, D. (2017). "The world's first trees grew by splitting their guts." *Science*, 23 October 2017. https://www.science.org/content/article/world-s-first-trees-grew-splitting-their-guts

Ogden, J., Wilson, A., Hendy, C., Newnham, R. M. and Hogg, A. G. (1992). "The Late Quaternary History of Kauri (*Agathis australis*) in New Zealand and Its Climatic Significance." *Journal of Biogeography*, vol. 19, issue 6, pp. 611–22.

Ortigosa, F., Lobato-Fernández, C., Shikano, H., Ávila, C., Taira, S., Cánovas, F.M. et al. (2022). "Ammonium regulates the development of pine roots through hormonal crosstalk and differential expression of transcription factors in the apex." *Plant, Cell & Environment*, vol. 45, pp. 915–35.

Orwell, G. (1938). *Homage to Catalonia* (Secker & Warburg, London)

Pauli, J. N. et al. (2014). "A syndrome of mutualism reinforces the lifestyle of a sloth." *Proceedings of the Royal Society B: Biological Sciences*, vol. 281, no. 1778.

Pausas, J. G. and Keeley, J. E. (2009). "A Burning Story: The Role of Fire in the History of Life." *BioScience*, vol. 59, pp. 593–601.

Pellmyr, O., (2003). "Yuccas, Yucca Moths, and Coevolution: A Review." *Annals of the Missouri Botanical Garden*, vol. 90, no. 1, pp. 35–55.

Pennisi, E. (2023). "Wandering Seeds." *Science*, vol. 381, issue 6658, pp. 598–601.

Pereira, C. C., Fernandes, G. W. and Cornelissen, T. (2022). "A double defensive mutualism? A case between plants, extrafloral nectaries, and trophobionts." *Alpine Entomology*, vol. 6, pp. 129–31.

Philippe, M. (2011). "Combien d'espèces d'Araucarioxylon?" *Palevol*, vol. 10, issues 2–3, pp. 201–8.

Pollan, M. (2001). *The Botany of Desire: A Plant's-Eye View of the World* (Random House, New York)

— (1991). *Second Nature: A Gardener's Education* (Grove Atlantic, New York)

Popkin, G. (2022). "Are Trees Talking Underground? For Scientists, It's in Dispute." *New York Times*, 7 November 2022. https://www.nytimes.com/2022/11/07/science/trees-fungi-talking.html

Powers, R. (2018). *The Overstory* (Norton, New York)

Prance, G. (2023). *The Amazon Forest and Its People in Black and White* (Butterflies and Amazonia Books)

— (2015). *Flowers, Fruits and Fables of Amazonia* (Butterflies and Amazonia Books)

— (2021). *Out of the Amazon* (Butterflies and Amazonia Books)

Priyashantha, A. K. H., Dai, D.-Q., Bhat, D. J., Stephenson, S. L., Promputtha, I., Kaushik, P., Tibpromma, S. and Karunarathna, S. C. (2023). "Plant–Fungi Interactions: Where It Goes?" *Biology*, vol. 12, issue 6, 809.

Proffit, M., et al. (2008). "Signaling receptivity: Comparison of the emission of volatile compounds by figs of *Ficus hispida* before, during and after the phase of receptivity to pollinators." *Symbiosis*, vol. 45, pp. 15–24.

Pyne, S. J. (1991). *Burning Bush: A Fires History of Australia* (Holt, New York)

Rackham, O. (1986). *The History of the Countryside* (Weidenfeld & Nicolson, London)

— (1989). *The Last Forest: The Story of Hatfield Forest* (J. M. Dent & Sons, London)

— (2006). *Woodlands*, Collins New Naturalist Library, vol. 100 (William Collins, London)

Ramage, B. S., O'Hara, K. L. and Caldwell, B. T. (2010). "The role of fire in the competitive dynamics of coast redwood forests." *Ecosphere*, vol. 1, issue 6.

Rasheed, M. U., Brosset, A. and Blande, J. D. (2023). "Tree Communication: The Effects of 'Wired' and 'Wireless' Channels on Interactions with Herbivores." *Current Forestry Reports*, vol. 9, pp. 33–47.

Raven, J. A. and Andrews M. (2010). "Evolution of tree nutrition." *Tree Physiology*, vol. 30, pp. 1050–71.

Read, J. et al. (2021). "Population age structures, persistence and flowering cues in *Cerberiopsis candelabra* (Apocynaceae), a long-lived monocarpic rain-forest tree in New Caledonia." *Journal of Tropical Ecology*, vol. 37, issue 6, pp. 263–75.

Reichert, T., Rammig, A., Fuchslueger, L., Lugli, L., Quesada, C. and Fleischer, K. (2009). "Tansley review: Plant phosphorus-use and -acquisition strategies in Amazonia." *New Phytologist*, vol. 234, pp. 1126–43.

Ren, X., Zhang, G., Chen, Z. and Zhu, J. (2023). "The Influence of Wind-Induced Response in Urban Trees on the Surrounding Flow Field." *Atmosphere*, vol. 14, issue 6, 1010.

Rheinische Zeitung (1842). "Proceedings of the Sixth Rhine Province Assembly. Third Article. Debates on the Law on Thefts of Wood," trans. Dutt, C. First published in the supplement to the *Rheinische Zeitung*, nos. 298, 300, 303, 305 and 307, 25, 27 and 30 October, 1 and 3 November 1842.

Ribeiro, C. M. and Cardoso, E. J. B. N. (2012). "Isolation, selection and characterization of root-associated growth promoting bacteria in Brazil pine (*Araucaria angustifolia*)." *Microbiological Research*, vol. 167, issue 2, pp. 69–78.

Ribeiro, J. et al. (1999). *Flora Da Reserva Ducke: Guia de Identificação das Plantas Vasculares de Uma Floresta de Terra-Firme Na Amazônia Central.* (INPA-DFID, Manaus, Amazonas)

Rich, M. K., Nouri, E., Courty, P.-E. and Reinhardt, D. (2017). "Diet of Arbuscular Mycorrhizal Fungi: Bread and Butter?" *Trends in Plant Science*, vol. 22, issue 8, pp. 652–60.

Riches, M., Berg, T. C., Vermeuel, M. P., Millet, D. B. and Farmer, D. K. (2024). "Wildfire smoke directly changes biogenic volatile organic emissions and photosynthesis of ponderosa pines." *Geophysical Research Letters*, vol. 51, issue 6.

Riddle, S. (2016). "How bees see and why it matters." *Bee Culture*, 20 May 2016. https://beeculture.com/bees-see-matters/

Roberts, P. (2021). *Jungle: How Tropical Forests Shaped World History—and Us* (Viking Press, London)

Robin, M., Römermann, C., Niinemets, Ü. et al. (2024). "Interactions between leaf phenological type and functional traits drive variation in isoprene emissions in central Amazon forest trees." *Frontiers in Plant Science*, vol. 15, 1522606.

Roche, J. R., Mitchell, F. J. G. and Waldren, S. (2009). "Plant community ecology of *Pinus sylvestris*, an extirpated species reintroduced to Ireland." *Biodiversity and Conservation*, vol. 18, 2185.

Rowntree, L. B. et al. (2014). "Afforestation, Fire, and Vegetation Management in the East Bay Hills of the San Francisco Bay Area." *Yearbook of the Association of Pacific Coast Geographers*, vol. 56.

Royer, A. M., Streisfeld, M. A. and Smith, C. I. (2016). "Population genomics of divergence within an obligate pollination mutualism: Selection maintains differences between Joshua tree species." *American Journal of Botany*, vol. 103, issue 10, pp. 1730–41.

Sample, I. (2016). "Family tree fall: human ancestor Lucy died in arboreal accident, say scientists." *Guardian*, 29 August 2016. https://www.theguardian.com/science/2016/aug/29/family-tree-fall-human-ancestor-lucy-died-in-arboreal-accident-say-scientists

Sánchez-Falfan, A., Esperón-Rodríguez, M., Cervantes-Pérez, J., Ballinas, M. and Barradas, V. L. (2023). "How Important Are Fog and the Cloud Forest as a Water Supply in Eastern Mexico?" *Water*, vol. 15, issue 7, 1286.

Sandford, S. (2022). *Vegetal Sex: Philosophy of Plants* (Bloomsbury, London)

Santana, M. A. E. and Okino, E. Y. A. (2007). "Chemical composition of 36 Brazilian Amazon forest wood species." *Holzforschung*, vol. 61, no. 5, pp. 469–77.

Santana, M. A., Pihakaski-Maunsbach, K., Sandal, N., Marcker, K. A. and Smith, A. G. (1998). "Evidence that the plant host synthesizes the heme moiety of leghemoglobin in root nodules." *Plant Physiology*, vol. 116, issue 4, pp. 1259–69.

Satyal, P., Paudel, P., Raut, J., Deo, A., Dosoky, N. S. and Setzer, W. N. (2013). "Volatile constituents of *Pinus roxburghii* from Nepal." *Pharmacognosy Research*, vol. 5, issue 1, pp. 43–8.

Sazima, M., Buzato, S. and Sazima, I. (1999). "Bat-pollinated Flower Assemblages and Bat Visitors at Two Atlantic Forest Sites in Brazil." *Annals of Botany*, vol. 83, pp. 705–12.

Schlanger, Z. (2024). *The Light Eaters: How the Unseen World of Plant Intelligence Offers a New Understanding of Life on Earth* (HarperCollins, London)

Scott, J. C. (2017). *Against the Grain: A Deep History of the Earliest States* (Yale University Press, New Haven)

Seyfullah, L. J. (2023). "Amber and the Cretaceous Resinous Interval," *Earth-Science Reviews*, vol. 243.

Seyfullah, L. J. et al. (2018). "Production and preservation of resins—past and present." *Biological Reviews*, vol. 93, issue 3, pp. 1684–1714.

Shear, W. A., Bonamo, P. M., Grierson J. D., Rolfe, W. D. I., Smith, E. L. and Norton, R. A. (1983). "Early Land Animals in North America: Evidence from Devonian Age Arthropods from Gilboa, New York." *Science*, vol. 224, issue 4648, pp. 492–4.

Sheil, D. and Murdiyarso, D. (2009). "How Forests Attract Rain: An Examination of a New Hypothesis." *BioScience*, vol. 59, pp. 341–7.

Sheldrake, M. (2020). *Entangled Life: How Fungi Make Our Worlds, Change Our Minds and Shape Our Futures.* (Bodley Head, London)

Shemesh, H., Boaz, B. E., Millar, C. I. and Bruns, T. B. (2020). "Symbiotic interactions above treeline of long-lived pines:

Mycorrhizal advantage of limber pine (*Pinus flexilis*) over Great Basin bristlecone pine (*Pinus longaeva*) at the seedling stage." *Journal of Ecology*, vol. 108, pp. 908–16.

Shigo, A. L. (1985). "Compartmentalization of Decay in Trees." *Scientific American*, vol. 252, no. 4, pp. 96–105.

Shilling, J. E. (2018). "Aircraft observations of the chemical composition and aging of aerosol in the Manaus urban plume during GoAmazon 2014/5." *Atmospheric Chemistry and Physics*, vol. 18, issue 14.

Shkolnik, D., Krieger, G., Nuriel, R. and Fromm, H. (2016). "Hydrotropism: Root Bending Does Not Require Auxin Redistribution," Letter to the Editor. *Molecular Plant*, vol. 9, issue 5, pp. 757–9.

Shrivastava, M., Andreae, M. O., Artaxo, P. et al. (2019). "Urban pollution greatly enhances formation of natural aerosols over the Amazon rainforest." *Nature Communications*, vol. 10, issue 1.

Simard, S. (2021). *Finding the Mother Tree: Discovering the Wisdom of the Forest* (Allen Lane, London)

Simard, S., Perry, D. A., Jones, M. D., Myrold, D. D., Durall, D. M. and Molina, R. (1997). "Net Transfer of Carbon Between Ectomycorrhizal Tree Species in the Field." *Nature*, vol. 388, pp. 579–82.

Simon, L., Bousquet, J., Lévesque, R. et al. (1993). "Origin and diversification of endomycorrhizal fungi and coincidence with vascular land plants." *Nature*, vol. 363, pp. 67–9.

Sivaramakrishnan, K. (1996). "The Politics of Fire and Forest Regeneration in Colonial Bengal." *Environment and History*, vol. 2, no. 2, pp. 145–194.

Sleeman, A., McConnell, B. and Gatley, S. (2004). *Understanding Earth Processes, Rocks and the Geological History of Ireland.* Geological Survey of Ireland. https://www.gsi.ie/documents/UnderstandingEarth_bookmarked.pdf

Smart, M. S., Filippelli, G., Gilhooly III, W. P., Marshall, J. E. A. and Whiteside, J. H. (2022). "Enhanced terrestrial nutrient release during the Devonian emergence and expansion of forests: Evidence from lacustrine phosphorus and geochemical records." *GSA Bulletin*, vol. 135, issues 7–8, pp. 1879–98.

Smith, C., Baker, J. C. A. and Spracklen, D. V. (2023). "Tropical deforestation causes large reductions in observed precipitation." *Nature*, vol. 615, pp. 270–5.

Smith, G. R., Finlay, R. D., Stenlid, J., Vasaitis, R. and Menkis, A. (2017). "Growing evidence for facultative biotrophy in saprotrophic fungi: data from microcosm tests with 201 species of wood-decay basidiomycetes." *New Phytologist*, vol. 215, pp. 747–55.

Smith, M. N., Taylor, T. C., van Haren, J. et al. (2020). "Empirical evidence for resilience of tropical forest photosynthesis in a warmer world." *Nature Plants*, vol. 6, pp. 1225–30.

Smith, S. D. and Rausher, M. D. (2011). "Gene Loss and Parallel Evolution Contribute to Species Difference in Flower Color." *Molecular Biology and Evolution*, vol. 28, pp. 10, pp. 2799–810.

Smith, S. M. (2014). "What are strigolactones and why are they important to plants and soil microbes?" *BMC Biology*, vol. 12, 19.

Solnit, R, (2021). *Orwell's Roses* (Viking Press, New York)

Som, T. and Dobson, K. (2024). " 'I struggled with anthropomorphisms': On the Problem of Metaphors, Happiness, and Forests in *Finding the Mother Tree*." *ISLE: Interdisciplinary Studies in Literature and Environment*, isae048.

Staub, F. (1995), "Dodo and Solitaires, Myth and Reality." Potomitan. https://www.potomitan.info/dodo/c32.php

Steidinger, B. S., Crowther, T. W., Liang, J. et al. (2019). "Climatic controls of decomposition drive the global biogeography of forest-tree symbioses." *Nature*, vol. 569, pp. 404–8.

Stein, W. E. et al. (2020). "Mid-Devonian *Archaeopteris* Roots Signal Revolutionary Change in Earliest Fossil Forests." *Current Biology*, vol. 30, issue 3, pp. 421–31.

Stephens, S. L., Martin, R. E. and Clinton, N. E. (2007). "Prehistoric fire area and emissions from California's forests, woodlands, shrublands, and grasslands." *Forest Ecology and Management*, vol. 251, issue 3.

Steward, G. A. and Beveridge, A. E. (2010). "A review of New Zealand kauri (*Agathis australis* (D.Don) Lindl.): its ecology, history, growth and potential for management for timber." *New Zealand Journal of Forestry Science*, vol. 40, pp. 33–59.

Stewart, G. (1948). *Fire* (Randon House, New York)

Stewart, W. N. and Rothwell, G. W. (1993). *Paleobotany and the Evolution of Plants*, 2nd ed. (Cambridge University Press, Cambridge)

Strullu-Derrien, C., Selosse, M.-A., Kenrick, P. and Martin, F.M. (2018). "The origin and evolution of mycorrhizal symbioses: from paleomycology to phylogenomics." *New Phytologist*, vol. 220, pp. 1012–30.

Su, C. et al. (2023). "Tree architecture: A strigolactone-deficient mutant reveals a connection between branching order and auxin gradient along the tree stem." *Proceedings of the National Academy of Sciences of the USA*, vol. 120, no. 48.

Su, S.-H., Gibbs, N. M., Jancewicz, A. L., Masson, P. H. (2017). "Molecular Mechanisms of Root Gravitropism," *Current Biology*, vol. 27, issue 17, pp. R964–R972.

Svensson, G. P., Raguso, R. A., Flatz, R. and Smith, C. I. (2016). "Floral scent of Joshua trees (*Yucca brevifolia sensu lato*): Divergence in scent profiles between species but breakdown of signal integrity in a narrow hybrid zone." *American Journal of Botany*, vol. 103, issue 10, pp. 1793–1802.

Takahashi, T. et al. (2014). "Identification of Phenylpropanoids in Fig (*Ficus carica* L.) Leaves." *Journal of Agricultural and Food Chemistry*, vol. 62, issue 41, pp. 10076–83.

Tavares, J. P. N. (2011). "Interaction between vegetation and the atmosphere in cloud and rain formation in the Amazon: A review." *Estudos Avançados*, vol. 26, issue 74, pp. 219–28.

Taylor, C. and Dewsbury, B. M. (2017). "On the Problem and Promise of Metaphor Use in Science and Science Communication." *Journal and Biology Evolution*, vol. 19, issue 1, pp. 1–5.

Thomas, B. R. et al. (1966). "The Chemistry of the Order Araucariales Part 4. The Bled Resins of *Agathis australis*." *Acta Chemica Scandinavica*, vol. 20, pp. 1074–81.

Thomas, P. (2022). *Trees*, Collins New Naturalist Library, vol. 11 (William Collins, London)

Thorpe, S. K. and Crompton, R. H. (2006). "Orangutan positional behavior and the nature of arboreal locomotion in

Hominoidea." *American Journal of Physical Anthropology*, vol. 131, issue 3, pp. 384–401.

Tomlinson, P. (2016). *The Botany of Mangroves*, 2nd ed. (Cambridge University Press, Cambridge)

Trickey, E. (2016). "The Story Behind a Forgotten Symbol of the American Revolution: The Liberty Tree." *Smithsonian Magazine*. https://www.smithsonianmag.com/history/story-behind -forgotten-symbol-american-revolution-liberty-tree-180959162/

Urrutia-Jalabert, R., Malhi, Y. and Lara, A. (2015). "The Oldest, Slowest Rainforests in the World? Massive Biomass and Slow Carbon Dynamics of *Fitzroya cupressoides* Temperate Forests in Southern Chile." *PLoS ONE*, vol. 10, issue 9.

Valderrama-Martín, J. M., Ortigosa, F., Cantón, F. R. et al. (2024). "Emerging insights into nitrogen assimilation in gymnosperms." *Trees*, vol. 38, pp. 273–86.

Vamosi, J. C. and Vamosi, S. M. (2011). "Factors Influencing Diversification in Angiosperms: At the Crossroads of Intrinsic and Extrinsic Traits." *American Journal of Botany*, vol. 98, issue 3, pp. 460–71.

Van Gogh, V. (2020). *Vincent Van Gogh: A Life in Letters*, eds. Bakker, N., Jansen, L., Luijten, H. (Thames and Hudson, London)

Veilleux, C. C. et al. (2023). "Human subsistence and signatures of selection on chemosensory genes." *Communications Biology*, vol. 6, 683.

Vincke, C. and Thiry, Y. (2008). "Water table is a relevant source for water uptake by a Scots pine (*Pinus sylvestris* L.) stand: Evidences from continuous evapotranspiration and water table monitoring." *Agricultural and Forest Meteorology*, vol, 148, issue 10, pp. 1419–32.

Viney, M., Hickey, R. D. and Mustoe, G. E. (2019). "A Silicified Carboniferous Lycopsid Forest in the Colorado Rocky Mountains, USA." *Geosciences*, vol. 9, issue 12.

Vonlanthen, B. (2010). "Establishment and development of phreatophytic vegetation in the foreland of river oases at the southern rim of the Taklamakan Desert." PhD thesis. https://repo.bibliothek .uni-halle.de/bitstream/1981185920/6758/1/Establishment%20 and%20development%20of%20phreatophytic%20vegetation%20

in%20the%20foreland%20of%20river%20oases%20at%20the%20
southern%20rim%20of%20the%20Taklamakan%20Desert.pdf

Voosen, P. (2021). "Ancient kauri trees capture last collapse of Earth's magnetic field." *Science*. https://www.science.org /content/article/ancient-kauri-trees-capture-last-collapse -earth-s-magnetic-field

Wallace, A. R. (1853). *Palm Trees of the Amazon and Their Uses* (John Van Voorst, London)

Wang, A. Y., Peng, Y.-Q., Harder, L. D., et al. (2019). "The nature of interspecific interactions and codiversification patterns, as illustrated by the fig microcosm." *New Phytologist*, vol. 224, pp. 1304–15.

Wang, D. et al. (2019). "The Most Extensive Devonian Fossil Forest with Small Lycopsid Trees Bearing the Earliest Stigmarian Roots." *Current Biology*, vol. 29, pp. 2604–15.

Wang, J. et al. (2021). "Ancient noeggerathialean reveals the seed plant sister group diversified alongside the primary seed plant radiation." *Proceedings of the National Academy of Sciences of the USA*, vol. 118, no. 11.

Waters, M. T., Gutjahr, C., Bennett, T. and Nelson, D. C. (2017). "Strigolactone Signaling and Evolution." *Annual Review of Plant Biology*, vol. 68, pp. 291–322.

Weinhold, A., Döll, S., Liu, M., Schedl, A., Pöschl, Y., Xu, X., Neumann, S. and van Dam, N. M. (2022). "Tree species richness differentially affects the chemical composition of leaves, roots and root exudates in four subtropical tree species." *Journal of Ecology*, vol. 110, pp. 97–116.

Wengrow, D. and Graeber, D. (2021). *The Dawn of Everything: A New History of Humanity* (Allen Lane, London)

Whippo, C. W. and Hangarter, R. P. (2009). "The 'Sensational' Power of Movement in Plants: A Darwinian System for Studying the Evolution of Behavior." *American Journal of Botany*, vol. 96, issue 12, pp. 2115–27.

Whiteside J. H., Lindström, S., Irmis, R. B., Glasspool, I. J., Schaller, M. F., Dunlavey, M., Nesbitt, S. J., Smith, N. D. and Turner, A. H. (2015). "Extreme ecosystem instability suppressed tropical dinosaur dominance for 30 million years." *Proceedings of the National Academy of Sciences of the USA*, vol. 112, no. 26, pp. 7909–13.

Willis, K. (2024). *Good Nature: The New Science of How Nature Improves Our Health* (Bloomsbury, London)

Wohlleben, P. (2016). *The Hidden Life of Trees* (Greystone, Vancouver)

Wolfe, J. A. (1997). "Relations of Environmental Change to Angiosperm Evolution During the Late Cretaceous and Tertiary" in eds. Iwatsuki, K. and Raven, P. H., *Evolution and Diversification of Land Plants* (Springer, Tokyo)

Woodward, F. I. (1987). "Stomatal numbers are sensitive to increases in CO_2 from pre-industrial levels." *Nature*, vol. 327, issue 6123, pp. 617–18.

Woolf, V. (1925). *Mrs Dalloway* (Hogarth Press, London)

Wright, G. A. and Schiestl, F. (2009). "The evolution of floral scent: the influence of olfactory learning by insect pollinators on the honest signaling of floral rewards." *Functional Ecology*, vol. 23, pp. 841–51.

Yoneyama, K. (2019). "How Do Strigolactones Ameliorate Nutrient Deficiencies in Plants?" *Cold Spring Harbor Perspectives in Biology*, vol. 11, issue 8.

Yoneyama, K., Xie, X., Kim, H. I. et al (2012). "How do nitrogen and phosphorus deficiencies affect strigolactone production and exudation?" *Planta*, vol. 235, issue 6, pp. 1197–207.

Young, S. N. R. and Lundgren, M. R. (2023). "C4 photosynthesis in Paulownia? A case of inaccurate citations." *Plants, People, Planet*, vol. 5, issue 2, pp. 292–303.

Zarrillo, S., Gaikwad, N., Lanaud, C., et al. (2018). "The use and domestication of *Theobroma cacao* during the mid-Holocene in the upper Amazon." *Nature Ecology & Evolution*, vol. 2, issue 12, pp. 1879–88.

Zhang, Y., Xiao, G., Wang, X., Zhang, X. and Friml, J. (2019). "Evolution of fast root gravitropism in seed plants." *Nature Communications*, vol. 10, 3480.

Zuntini, A. R. et al. (2024). "Phylogenomics and the rise of the angiosperms." *Nature*, vol. 629, pp. 843–50.

Table of Geological Time

MILLIONS OF YEARS BP	ERA	PERIOD	SOME EVENTS DESCRIBED IN THIS BOOK
2500	PRE-CAMBRIAN	Proterozoic	First life begins. Algae and the beginnings of photosynthesis
570	LOWER PALEOZOIC	Cambrian	Trilobites
510	LOWER PALEOZOIC	Ordovician	Snowball frozen earth
438	LOWER PALEOZOIC	Silurian	Prototaxites, the first land plants
410	UPPER PALEOZOIC	Devonian	*Archeopteris* and the first trees
355	UPPER PALEOZOIC	Carboniferous	Coal formation; lycopsids and tree ferns
290	UPPER PALEOZOIC	Permian	Cycads, supercontinent Pangaea
250	MESOZOIC	Triassic	Gymnosperms (e.g., *Araucaria*)
205	MESOZOIC	Jurassic	Dinosaurs, *Sequoia*, *Ginkgo*
138	MESOZOIC	Cretaceous	Rise of the angiosperms
Asteroid lands at Chicxulub circa 66 million years ago			
65	TERTIARY	Palaeocene	Great flowering and domination of the angiosperms in widespread tropical and subtropical forest e.g. Winteraceae Liliaceae, *Nypa*, Magnoliaceae
53	TERTIARY	Eocene	Great flowering and domination of the angiosperms in widespread tropical and subtropical forest e.g. Winteraceae Liliaceae, *Nypa*, Magnoliaceae
36.5	TERTIARY	Oligocene	Cooling and drying leads to patchy forest; beeches, pines and cashews are common
23	TERTIARY	Miocene	American white oaks cross back into Europe; hornbeams and maples are common
5.3	TERTIARY	Pliocene	Temperate forest in the northern hemisphere
1.6	TODAY	Pleistocene	Glacial cycles
0.01	TODAY	Holocene	Humans develop; extinction of most megafauna

List of Illustrations

All images belong to the author, Harriet Rix, and/or her father, Martyn Rix, unless otherwise stated.

Acknowledgments

Trees grow across the world on one another's shoulders, and I clambered across the places I have written about here with help from many tree people, some in the form of books and articles and some in person. They form a great shadow-forest of mutual interest and support on every continent.

In Iraq my profound thanks go to the staff of Hasar Organisation, Hawkar Ali Abdulhaq, Mam Ali, Professor Saman Abdulrahman, Dr. Sajad Zalzala, and Tahir Salih Alkay. In the United States, the trustees and staff of the New York Botanic Garden—particularly Todd Forrest, William Stein, and the team at the New York State Museum in Albany, and the US Fire Administration and Cal Fire were superbly generous with time and expertise. In Pakistan, members of the Endowment Fund Trust for Preservation of the Heritage of Sindh and the Balochistan Forests and Wildlife Department, particularly Dr. Niaz Khan Kakar, as well as Siraj and Maqsood Ul-Mulk and their households who offered magnificent insights as well as fabulous hospitality; Gül Kalash explained many aspects of Kalash culture with patience and understanding. In Turkey the Güner family, Professor Dr. Necmi Aksoy, Dr. Elif Deniz Ülker, and Professor Dr. K. Husnu Can Baser provided invaluable advice. In Greece, Michael and Despina Moschos, Yiannis Chaldoupis, and Kalliope Stara corrected me in many ways. In Brazil unparalleled gratitude to Giovanna Claro Baré and her family, and thanks to Mike Hopkins, Curador do Herbário INPA, Christabel King, Professor Sir Ghillean Prance, James Willoughby, and Laurie Blair for generously sharing their knowledge, and in Belize Jaime Awe and the Belize Valley Archeological Reconnaissance Project, Ella Baron at the Ian Anderson Botanical Garden, and Bruce Holst at the Marie Selby Botanical Garden all provided invaluable help.

I owe thanks also to a huge number of friends who have traveled with and hosted me around the world. Yad Bayad Abdulqader has been an intrepid friend and companion on many tree-hunting adventures and I would like to thank his family for their hospitality, as well as their guidance, intelligence, poetic flair, and cosmopolitan wit. Lara Prior-Palmer, Jem Lowther, Emma Rix, Ahmad Ibrahim Qader, Benedikt Weiss, Ed Picton-Turbeville, and Josh Barley were not only superb travel companions but also read parts of the book and offered essential feedback. James was there at the beginning and the end. Robin Lane Fox taught me that big ideas are worth getting involved in, and that the idea shared is the idea enhanced, enriched, and brought to life. Mark Baczoni, Siepsah Tamoukh, Fergus McIntosh, Eli Berger, Özge Dursun, Dr. Jonah Rosenberg, Dr. Geri Della Rocca de Candal, Mor Timothius Mousa Alshamany, Abu Yousif and the monks of Dayro d-Mor Mattai, Gregory Long, Scott Newman, Alex and Eliza Chisholm, Cecilia Stinton, and Peter Treherne all gave me places to stay and write. Charlotte Hall, Tianyi Yu, Sophie Yeo, Dr. Meseret Oldjira, Dr. Thomas Probert, John d'Arcy, Mehul Srivastava, Lyndsie Bourgon, Professor Andrew Smith, Dr. Manuel Arroyo-Kalin, Joy Lo Dico, Keith Rushforth, Jess Hao, Jim McCue, Mikonori Ogisu, Professor Nick Lane, Maia Hrushkova, Ed Fortes, Masumi Yamanaka, Vir Bulchandani, and many others provided essential insights. The Tree Council, in particular Jon Stokes, the International Oak Society, the International Dendrological Society, the members of Hedgelink, and many tree officers up and down the UK taught me to look at trees in a way which I hope is reflected throughout the book. Thank you.

My Oxford tutor Professor Stuart Ferguson died last spring as I was finishing this book. He was far too young, and his patience in teaching the physical and philosophical aspects of bioenergetics has enriched my life hugely. Even younger, and also sadly missed, is Dr. Chris Scanlan, whose brilliance in immunology was only matched by his kindness. Both were true scientists who would be appalled by the mistakes in this book. The mistakes are all my own, but there would be many more without the help and guidance of the staff at the Herbarium, Library, and Jodrell Laboratory of the Royal

Botanical Gardens, Kew, The Linnaean Society, the Royal Society, and The London Library.

I was extremely lucky to be represented by Doug Young—wizard among agents—and all the staff at PEW. My parents read every page of this manuscript, twice or more. Thank you.

James Nightingale, Sally Sargeant, and Alex Bell copyedited, proofread, and indexed with consummate skill. Finally, Stuart Williams, editor of this book, offered unrivaled patience, tact, and technical skill in making the apparently impossible possible; to him, Laura Reeves, Rowena Skelton-Wallace, Joe Pickering, Annie Rose, and all the team at Bodley Head I owe some apologies, but most of all massive thanks.

Index

About the Author

Harriet Rix is a tree science consultant and writer. She was formerly based at the Tree Council, where she researched tree diseases and urban tree strategies. Before joining the tree sector, she worked in landmine clearance in the Middle East. She acted as scientific advisor on Adrian Grenier's climate documentary, was secretary of Hedge-link, and is a trustee of the Iraqi environmental charity Hasar.

Rix holds a biochemistry degree from the University of Oxford and a master's in the history and philosophy of science from the University of Cambridge. She was a London Library Emerging Writer, and her writing and photography have been published in the *Financial Times*, *London Review of Books*, *The Times Literary Supplement*, and more. *The Genius of Trees* is her first book.